Kristen Nehemiah Horst (Ed.)

Foreign Grain Beetle

AF694305

Kristen Nehemiah Horst (Ed.)

Foreign Grain Beetle

Beetle, Oryzaephilus surinamensis, Drosophila melanogaster, Flea, Louse

Dign Press

Imprint

Permission is granted to copy, distribute and/or modify this document under the terms of the GNU Free Documentation License, Version 1.2 or any later version published by the Free Software Foundation; with no Invariant Sections, with the Front-Cover Texts, and with the Back- Cover Texts. A copy of the license is included in the section entitled "GNU Free Documentation License".

All parts of this book are extracted from Wikipedia, the free encyclopedia (www.wikipedia.org).

You can get detailed informations about the authors of this collection of articles at the end of this book. The editors (Ed.) of this book are no authors. They have not modified or extended the original texts.

Pictures published in this book can be under different licences than the GNU Free Documentation License. You can get detailed informations about the authors and licences of pictures at the end of this book.

The content of this book was generated collaboratively by volunteers. Please be advised that nothing found here has necessarily been reviewed by people with the expertise required to provide you with complete, accurate or reliable information. Some information in this book maybe misleading or wrong. The Publisher does not guarantee the validity of the information found here. If you need specific advice (f.e. in fields of medical, legal, financial, or risk management questions) please contact a professional who is licensed or knowledgeable in that area.

Any brand names and product names mentioned in this book are subject to trademark, brand or patent protection and are trademarks or registered trademarks of their respective holders. The use of brand names, product names, common names, trade names, product descriptions etc. even without a particular marking in this works is in no way to be construed to mean that such names may be regarded as unrestricted in respect of trademark and brand protection legislation and could thus be used by anyone.

Cover image: www.ingimage.com
Concerning the licence of the cover image please contact ingimage.

Publisher:
Dign Press is a trademark of
International Book Market Service Ltd., 17 Rue Meldrum, Beau Bassin, 1713-01 Mauritius
Email: info@bookmarketservice.com
Website: www.bookmarketservice.com

Published in 2011

Printed in: U.S.A., U.K., Germany. This book was not produced in Mauritius.

ISBN: 978-613-6-59410-1

Contents

Articles

References

Article Licenses

Foreign grain beetle

Ahasverus advena	
Scientific classification	
Kingdom:	Animalia
Phylum:	Arthropoda
Class:	Insecta
Order:	Coleoptera
Family:	Silvanidae
Genus:	*Ahasverus*
Species:	***A. advena***
Binomial name	
Ahasverus advena (Waltl, 1832)	

Ahasverus advena, also referred to as the *foreign grain beetle* belongs to the beetle family Silvanidae, as does its relative the sawtoothed grain beetle.

Identification

The **foreign grain beetle** is a rather small beetle, approximately 2 mm ($^1/_{12}$ in) in length. This beetle can be distinguished chiefly by slight projections or knobs on each front corner of the pronotum, and its club-shaped antennae. The larvae are worm-like, cream-colored and often reach a length of 3 mm before pupating into darker adults. Males and females are identical in appearance both as larvae and adults. Adults are a reddish brownish color (occasionally black). They are very similar to the saw-toothed grain beetle, but lack the "sawtoothed" projections on the pronotum. These small beetles are strong fliers, and are often confused with fruit flies or gnats. However, these beetles are hard shelled, and lack clear wings. Some people can mistake them for fleas or lice, but neither of those insects can fly.

Natural history

Distribution

The foreign grain beetle is found in both tropical and temperate regions, since the temperature for complete development lies between 20 to 35 °C.

Behavior

They usually become apparent in early summer to late fall when humidity levels are highest. These beetles can only survive if relative humidity exceeds 70%. Unlike their name suggests, the only things they eat are mold or fungi. They acquired their name because they are often found feeding on moldy grain. However, grain injury by this pest is not severe enough to cause noticeable economic loss. They can appear near bathtubs, sinks or any moist area. They can enter houses from the walls near pipes.

Life cycle

Adult females can begin laying eggs around 3–4 days after pupating to an adult female. Daily oviposition rates are 1-4 eggs, but can be as high as 8–12 eggs. These eggs are laid singly or in clusters of two or three, and hatch in 4–5 days. The larval stage is completed in 11–19 days, and then it will pupate for 3–5 days. Under laboratory conditions, and a temperature of 27°C, the transition from egg to adult beetle takes around 3 weeks. Larval development times increase with decreasing humidity. Mated males and females have an average life span of 159 and 208 days, respectively. Unmated beetles live considerable longer, with a life span of about 275 days for males, and 300 days for females.

Impact on humans

These are harmless insects that sometimes inhabit dwellings. They infest a number of products such as grains, cereal products, oilseeds, dried fruit, and spices. An infestation will occur if mold has grown on these products already due to a damp environment. Hence, these beetles are considered good indicators of damp storage conditions, and spoiled food.[1] They do not bite or damage wood, fabric or stored food. They only feed on the mold and fungi that has grown on them. They are especially problematic in new houses due to the mold and fungal growth on new lumber, or wet wall board . In homes, they can be confused with ticks, fruit flies and fleas.

Control

The best way to control them is to vacuum your house at least three times a week, and eliminate the damp source. New homes naturally dry out within the first few years, and drying times can be accelerated by increasing ventilation by means of fans or air conditioning. Aerosol sprays of grain insecticide containing pyrethrins can also be used, but must be applied frequently. However, these Grain-Protectant insecticides will not kill the insects inside the grain kernels.[1]

Biological control

Various insect predators and parasitic wasps can be used to attack this beetle. However, they are only effective when used in large numbers. Biological control is generally not used because in the U.S., the Food and Drug Administration (FDA) and food processors do not permit the use of live insects in raw grain.[1]

References

[1] Weaver, D. 2004. Foreign Grain Beetle- Stored Grain. High Plains IPM Guide, Cooperate effort of University of Wyoming, University of Nebraska, Colorado State University and Montana State University.

- Bousquet, Yves (1990). "Beetles associated with stored products in Canada.An identification guide" (http://home.cc.umanitoba.ca/~fieldspg/fields/beetles.pdf). Canadian Government Publishing Centre Supply and Services Canada. Retrieved 2009-02-04.

Further reading

- Potter, Mike (1993-02-01). "Foreign Grain Beetle" (http://www.ca.uky.edu/entomology/entfacts/ef610.asp). University of Kentucky College of Agriculture. Retrieved 2007-07-13.

External links

- *Foreign Grain Beetle* (http://www.extension.umn.edu/yardandgarden/ygbriefs/e603foreigngb.html) on University of Minnesota, Yard & Garden Brief
- *Foreign Grain Beetle* (http://www.ca.uky.edu/entomology/entfacts/ef610.asp) on University of Kentucky College of Agriculture
- *Foreign Grain Beetle* (http://agspsrv34.agric.wa.gov.au/ento/pestweb/Query1_1.idc?ID=906693402) on PestWeb, Government of Australia
- *Foreign Grain Beetle* (http://www.grainscanada.gc.ca/storage-entrepose/sip-irs/fgb-cg-eng.htm) on Canadian Grain Commission
- "Foreign Grain Beetle" (http://www.ent.iastate.edu/clinic/node/11) on Iowa State University, Department of Etymology

Beetle

Beetle Temporal range: Pennsylvanian – Recent	

Collage of species of Coleoptera circumposed on a white background. From top left to bottom right: female Golden stag beetle (*Lamprima aurata*), Rhinoceros beetle (*Xylotrupes gideon*), a species of *Amblytelus*, Cowboy beetle (*Chondropyga dorsalis*), and a Long nose weevil (*Rhinotia hemistictus*).	
Scientific classification	
Kingdom:	Animalia
Phylum:	Arthropoda
Class:	Insecta
Subclass:	Pterygota
Infraclass:	Neoptera
Superorder:	Endopterygota
Order:	**Coleoptera** Linnaeus, 1758
Suborders	
• Adephaga • Archostemata • Myxophaga • Polyphaga *See subgroups of the order Coleoptera*	

Coleoptera (/koʊliːˈɒptərə/) is an order of insects commonly called **beetles**. The word "coleoptera" is from the Greek κολεός, *koleos*, "sheath"; and πτερόν, *pteron*, "wing", thus "sheathed wing". Coleoptera contains more species than any other order, constituting almost 25% of all known life-forms.[1] About 40% of all described insect species are beetles (about 400,000 species[2]), and new species are discovered frequently. Some estimates put the total number of species, described and undescribed, at as high as 100 million, but 1 million is a more likely figure.[3] The largest taxonomic family, the Curculionidae (the weevils or snout beetles), also belongs to this order.

The diversity of beetles is very wide-ranging. They are found in almost all habitats, but are not known to occur in the sea or in the polar regions. They interact with their ecosystems in several ways. They often feed on plants and fungi, break down animal and plant debris, and eat other invertebrates. Some species are prey of various animals including birds and mammals. Certain species are agricultural pests, such as the Colorado potato beetle *Leptinotarsa decemlineata*, the boll weevil *Anthonomus grandis*, the red flour beetle *Tribolium castaneum*, and the mungbean or cowpea beetle *Callosobruchus maculatus*, while other species of beetles are important controls of agricultural pests. For example, beetles in the family Coccinellidae ("ladybirds" or "ladybugs") consume aphids, scale insects, thrips, and other plant-sucking insects that damage crops.

Species in the order Coleoptera are generally characterized by a particularly hard exoskeleton and hard forewings (elytra, singular elytron). These elytra distinguish beetles from most other insect species, except for a few species of Hemiptera. The beetle's exoskeleton is made up of numerous plates called sclerites, separated by thin sutures. This design creates the armored defenses of the beetle while maintaining flexibility. The general anatomy of a beetle is quite uniform, although specific organs and appendages may vary greatly in appearance and function between the many families in the order. Like all insects, beetles' bodies are divided into three sections: the head, the thorax, and the abdomen. Coleopteran internal morphology is similar to other insects, although there are several examples of novelty. Such examples include species of water beetle who use air bubbles in order to dive under the water, and can remain submerged thanks to passive diffusion allowing oxygen to transfer from the water into the bubble.

Beetles are endopterygotes, which means that they undergo complete metamorphosis, a biological process by which an animal physically develops after birth or hatching, undergoing a series of conspicuous and relatively abrupt change in the its body structure. Coleopteran species have an extremely intricate behavior when mating, using such methods as pheromones for communication to locate potential mates. Males may fight for females using very elongated mandibles, causing a strong divergence between males and females in sexual dimorphism.

Etymology

Coleoptera comes from the Greek *koleopteros*, literally "sheath-wing", from *koleos* meaning "sheath", and *pteron*, meaning "wing". The name was given to the group by Aristotle for their elytra, hardened shield-like forewings. The English name "beetle" comes from the Old English word *bitela*, literally meaning small biter, deriving from the word *bitel*, which means biting.[4] In addition to names including the word "beetle", individual species of Coleoptera have a variety of common names, including fireflies, June bugs, ladybugs and weevils.[1]

Distribution and diversity

Beetles are one of the largest orders of insects, with 350,000–400,000 species in four suborders (Adephaga, Archostemata, Myxophaga, and Polyphaga), making up about 40% of all insect species described. Even though classification at the family level is a bit unstable, there are about 500 recognized families and subfamilies.[1] [5] One of the first proposed estimates of the total number of beetle species on the planet, based on field data rather than on catalog numbers. The technique used for his original estimate, possibly as many as 12,000,000 species, was criticized, and was later revised, with estimates of 850,000–4,000,000 species proposed. Some 70–95% of all beetle species, depending on the estimate, remain undescribed. The beetle fauna is not equally well known in all parts of the world. For example, the known beetle diversity of Australia is estimated at 23,000 species in 3265 genera and 121 families. This is slightly lower than reported for North America, a land mass of similar size with 25,160 species in 3526 genera and 129 families. While other predictions show there could be as many as 28,000 species in North America, including those currently undescribed, a realistic estimate of the little-studied Australian beetle fauna's true diversity could vary from 80,000 to 100,000.[6]

Patterns of beetle diversity can be used to illustrate factors that have led to the success of the group as a whole. Based on estimates for all 165 families, more than 358,000 species of beetles have been described and are considered valid. Most species (about 62%) are in six extremely diverse families, each with at least 20,000 described species: Curculionidae, Staphylinidae, Chrysomelidae, Carabidae, Scarabaeidae, and Cerambycidae. The smaller families account for 22% of the total species - 127 families with fewer than 1000 described species and 29 families with 1000–6000 described species. So, the success of beetles as a whole is driven not only by several extremely diverse lineages, but also by a high number of moderately successful lineages. The patterns seen today indicate that beetles went through a massive adaptive radiation early in their evolutionary history, with many of the resulting lineages flourishing through hundreds of millions of years to the present. The adaptive radiation of angiosperms helped drive the diversification of beetles, as four of the six megadiverse families of beetles are primarily angiosperm-feeders:Curculionidae, Chrysomelidae, Scarabaeidae, and Cerambycidae. However, even without the phytophagous groups, lineages of predators, scavengers, and fungivores are tremendously successful.[6] Coleoptera are found in nearly all natural habitats, including freshwater and marine habitats, everywhere there is vegetative foliage, from trees and their bark to flowers, leaves, and underground near roots- even inside plants in galls, in every plant tissue including dead or decaying ones.[7]

External Morphology

Beetles are generally characterized by a particularly hard exoskeleton and hard forewings (elytra). The beetle's exoskeleton is made up of numerous plates called sclerites, separated by thin sutures. This design provides armored defenses while maintaining flexibility. The general anatomy of a beetle is quite uniform, although specific organs and appendages may vary greatly in appearance and function between the many families in the order. Like all insects, beetles' bodies are divided into three sections: the head, the thorax, and the abdomen.[1]

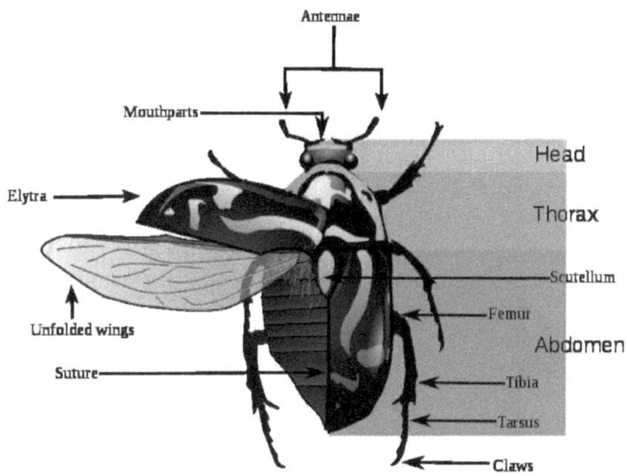

The morphology of a beetle, with a fiddler beetle as an example species

A beetle, just before takeoff. The elytra and unfolding wings can clearly be seen.

Head

Scarabaeus viettei (syn. *Madateuchus viettei*, Scarabaeidae) showing a "shovel head" adaptation.

The head, having mouthparts projecting forward or sometimes downturned, is usually heavily sclerotized and varies in size.[5] The eyes are compound and may display remarkable adaptability, as in the case of whirligig beetles (family Gyrinidae), where they are split to allow a view both above and below the waterline. Other species also have divided eyes – some longhorn beetles (family Cerambycidae) and weevils – while many have eyes that are notched to some degree. A few beetle genera also possess ocelli, which are small, simple eyes usually situated farther back on the head (on the vertex).

Beetles' antennae are primarily organs of smell, but may also be used to feel out a beetle's environment physically. They may also be used in some families during mating, or among a few beetles for defence. Antennae vary greatly in form within the Coleoptera, but are often similar within any given family. In some cases, males and females of the same species will have different antennal forms. Antennae may be clavate (flabellate and lamellate are sub-forms of clavate, or clubbed antennae), filiform, geniculate, moniliform, pectinate, or serrate.

Head of *Cephalota circumdata*, showing the compound eyes and mouthparts

Beetles have mouthparts similar to those of grasshoppers. Of these parts, the most commonly known are probably the mandibles, which appear as large pincers on the front of some beetles. The mandibles are a pair of hard, often tooth-like structures that move horizontally to grasp, crush, or cut food or enemies (see defence, below). Two pairs of finger-like appendages are found around the mouth in most beetles, serving to move food into the mouth. These are the maxillary and labial palpi. In many species the mandibles are sexually dimorphic, with the males' enlarged enormously compared with those of females of the same species.[5]

Thorax

The thorax is segmented into the two discernible parts, the pro- and pterathorax. The pterathorax is the fused meso- and metathorax, which are commonly separate in other insect species, although flexibly articulate from the prothorax. When viewed from below, the thorax is that part from which all three pairs of legs and both pairs of wings arise. The abdomen is everything posterior to the thorax.[1] When viewed from above, most beetles appear to have three clear sections, but this is deceptive: on the beetle's upper surface, the middle "section" is a hard plate called the pronotum, which is only the front part of the thorax; the back part of the thorax is concealed by the beetle's wings. This further segmentation is usually best seen on the abdomen.

Extremities

Acilius sulcatus, a diving beetle showing hind legs adapted for life in water

The multi-segmented legs end in two to five small segments called tarsi. Like many other insect orders beetles bear claws, usually one pair, on the end of the last tarsal segment of each leg. While most beetles use their legs for walking, legs may be variously modified and adapted for other uses. Among aquatic families – Dytiscidae, Haliplidae, many species of Hydrophilidae and others – the legs, most notably the last pair, are modified for swimming and often bear rows of long hairs to aid this purpose. Other beetles have fossorial legs that are widened and often spined for digging. Species with such adaptations are found among the scarabs, ground beetles, and clown beetles (family Histeridae). The hind legs of some beetles, such as flea beetles (within Chrysomelidae) and flea weevils (within Curculionidae), are enlarged and designed for jumping.

Wings

The elytra are connected to the pterathorax, so named because it is where the wings are connected (*pteron* meaning "wing" in Greek).[1] The elytra are not used for flight, but tend to cover the hind part of the body and protect the second pair of wings (*alae*). They must be raised in order to move the hind flight wings. A beetle's flight wings are crossed with veins and are folded after landing, often along these veins, and stored below the elytra. In some beetles, the ability to fly has been lost. These include some ground beetles (family Carabidae) and some "true weevils" (family Curculionidae), but also desert- and cave-dwelling species of other families. Many have the two elytra fused together, forming a solid shield over the abdomen. In a few families, both the ability to fly and the elytra have been lost, with the best known example being the glow-worms of the family Phengodidae, in which the females are larviform throughout their lives.

Abdomen

The abdomen is the section behind the metathorax, made up of a series of rings, each with a hole for breathing and respiration, called a spiracle; composing three different segmented sclerites: the tergum, pleura, and the sternum. The tergum in almost all species is membranous, or usually soft and concealed by the wings and elytra when not in flight. The pleura (singular: pleuron) are usually small or hidden in some species, with each pleuron having a single spiracle. The sternum is the most widely visible part of the abdomen, being a more or less scelortized segment. The abdomen itself does not have any appendages, however some species (for example,, *Mordellidae*) have articulating sternal lobes.[8]

Internal Morphology

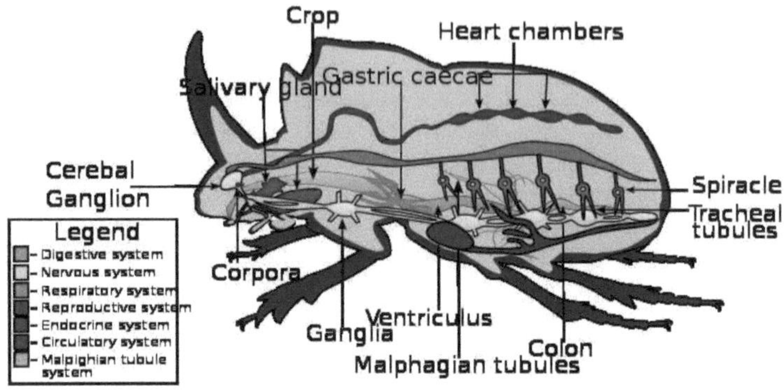

A diagram showing the general internal anatomy of species of coleoptera

Digestive system

The digestive system of beetles is primarily based on plants which they for the most part feed upon, with mostly the anterior midgut performing digestion, although in predatory species (for example Carabidae) most digestion occurs in the crop by means of midgut enzymes. In Elateridae species, the predatory larvae defecate enzymes on their prey, with digestion being extraorally.[1] The alimentary canal basically consists of a short narrow pharynx, a widened expansion, the crop and a poorly developed gizzard. After there is a midgut, that varies in dimensions between species, with a large amount of cecum, with a hindgut, with varying lengths. There are typically four to six Malpighian tubules.[5]

Nervous system

The nervous system in beetles contains all the types found in insects, varying between different species. With three thoracic and seven or eight abdominal ganglia can be distinguished to that in which all the thoracic and abdominal ganglia are fused to form a composite structure.[1]

Respiratory system

Oxygen is obtained via a tracheal system. Air enters a series of tubes along the body through openings called spiracles, and is then taken into increasingly finer fibers.[1] This bubble may be held under the elytra or it may be trapped against the body using specialized hairs. The bubble usually covers one or more spiracles so the insect can breathe air from the bubble while submerged. An air bubble provides an insect with only a short-term supply of oxygen, but thanks to its unique physical properties, oxygen will diffuse into the bubble and displacing the nitrogen, called passive diffusion, however the volume of the bubble eventually diminishes and the beetle will have to return to the surface.[9]

Circulatory system

Pumping movements of the body force the air through the system. Beetles have hemolymph instead of blood like other insect species, the open circulatory system of the beetle is driven by a tube-like heart attached to the top inside of the thorax. Some species of diving beetles (**Dytiscidae**) carry a bubble of air with them whenever they dive beneath the water surface.[1]

Specialized organs

Different glands specialize for different pheromones produced for finding mates. Pheromones from species of Rutelinea are produced from epithelial cells lining the inner surface of the apical abdominal segments or amino acid based pheromones of Melolonthinae from eversible glands on the abdominal apex. Other species produce different types of pheromones. Dermestids produce esters, and species of Elateridae produce fatty-acid-derived aldehydes and acetates.[1] For means of finding a mate also, fireflies (Lampyridae) utilized modified fat body cells woth transparent surfaces backed with reflective uric acid crystals to biosynthetically produce light, or bioluminescence. The light produce is highly efficient, as it is produced by oxidation of luciferin by the enzymes luciferase in the presence of ATP (adenosine triphospate) and oxygen, producing oxyluciferin, carbon dioxide, and light.[1]

A notable number of species have developed special glands that produce chemicals for deterring predators (see Defense and predation). The Ground beetle's (of Carabidae) defensive glands, located at the posterior, produce a variety of hydrocarbons, aldehydes, phenols, quinones, esters, and acids released from an opening at the end of the abdomen. While african carabid beetles (for example,, *Anthia* and *Thermophilium*) employ the same chemicals as ants: formic acid.[10] While Bombardier beetles have well-developed, like other carabid beetles, pygidial glands that empty from the lateral edges of the intersegment membranes between the seventh and eighth abdominal segments. The gland is made of two containing chambers. The first holds hydroquinones and hydrogen peroxide, with the second holding just hydrogen peroxide plus catalases. These chemicals mix and result in an explosive ejection, forming temperatures of around 100 C, with the breakdown of hydroquinone to H_2 + O_2 + quinone, with the O_2 propelling the excretion.[1]

Tympanal organs or hearing organs, which is a membrane (tympanum) stretched across a frame backed by an air sac and associated sensory neurons, are described in two families.[11] Several species of the genus *Cicindela* (**Cicindelidae**) have ears on the dorsal surface of the first abdominal segment beneath the wing; two tribes tribes in the family Dynastinae (**Scarabaeidae**) have ears just beneath the pronotal shield or neck membrane. The ears of both families are to ultrasonic frequencies, with strong evidence that they function to detect the presence of bats via there ultrasonic echolocation. Even though beetles constitute a large order and live in a variety of niches, examples of hearing is surprisingly lacking in species, though it is likely that most are just undiscovered.[1]

Reproduction and development

Beetles are members of Endopterygota, which means like most other insects under complete metamorphosis, which consists of four main stages: the egg, the larva, the pupa, and the imago or adult. The larvae are commonly called grubs and the pupa are called cocoons.

Mating

Beetles may display extremely intricate behavior when mating. Pheromone communication is likely to be important in the location of a mate. Different species use different chemicals for their pheromones. Some scarab beetles (for example,, Rutelinae) utilize pheromones derived from fatty acid synthesis, while other scarab beetles use amino acids and terpenoid compounds (for example,, Melolonthinae). Another way species of Coleoptera find mates is the use of biosynthesized light, or bioluminescence. This special form of a mating call is confined to fireflies (**Lampyridae**) by the use of abdominal light producing organs. The males and females engage in complex dialogue before mating, identifying different species by differences in duration, flight patterns, composition, and intensity.[1]

Punctate flower chafers (*Neorrhina punctata*, Scarabaeidae) mating

Before mating male and females may engage in various forms of behavior. Males and females may stridulate, or vibrate the object they are on. In some species (for example,, Meloidae) the male climbs onto the dorsum of the female and stroke his antennae on her head, palps and antennae. In the genus Eupompha of said family, the males draws the antennae along the longitudinal vertex on the male. They may not mate at all if they do not perform the precopulatory ritual.[1]

Conflict can play a part in the mating rituals of species such as burying beetles (genus *Nicrophorus*) where conflicts between males and females rage until only one of each is left, thus ensuring reproduction by the strongest and fittest. Many male beetles are territorial and will fiercely defend their small patch of territory from intruding males. In such species, the males may often have horns on the head and/or thorax, making their overall body lengths greater than those of the females, unlike most insects. Pairing is generally short but in some cases will last for several hours. During pairing sperm cells are transferred to the female to fertilize the egg.[5]

Life Cycle

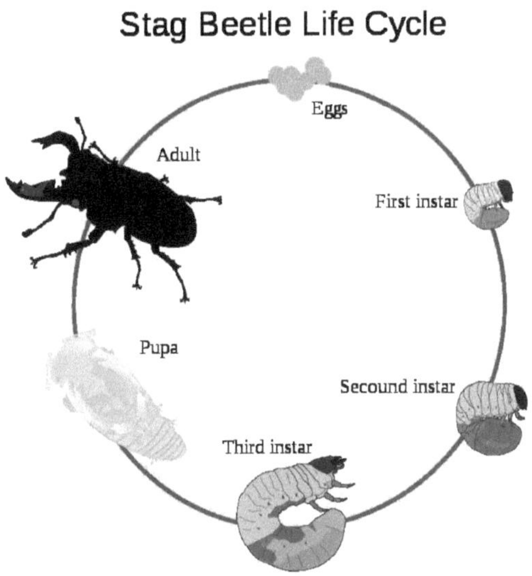

The life cycle of the stag beetle, including the 3 instars.

Egg

A single female may lay from several dozen to several thousand eggs during her lifetime. Eggs are usually laid according to the substrate the larva will feed on upon hatching. Among others, they can be laid loose in the substrate (for example, flour beetle), laid in clumps on leaves (for example, Colorado potato beetle), or individually attached (for example, mungbean beetle and other seed borers) or buried in the medium (for example, carrot weevil).

Parental care varies between species, ranging from the simple laying of eggs under a leaf to certain scarab beetles, which construct underground structures complete with a supply of dung to house and feed their young.[1] Other beetles are leaf rollers, biting sections of leaves to cause them to curl inwards, then laying their eggs, thus protected, inside.[1]

Larva

A scarabaeiform larva known as a Curl grub.

The larva is usually the principal feeding stage of the beetle life cycle. Larvae tend to feed voraciously once they emerge from their eggs. Some feed externally on plants, such as those of certain leaf beetles, while others feed within their food sources. Examples of internal feeders are most Buprestidae and longhorn beetles. The larvae of many beetle families are predatory like the adults (ground beetles, ladybirds, rove beetles). The larval period varies between species but can be as long as several years. The larva are highly varied amongst species, with a well-developed and sclerotized head and have distinguishable thoracic and abdominal segments (usually the tenth, though sometimes the eight or ninth).[5]

Beetle larvae can be differentiated from other insect larvae by their hardened, often darkened head, the presence of chewing mouthparts, and spiracles along the sides of the body. Like adult beetles, the larvae are varied in appearance, particularly between beetle families. Beetles whose larvae are somewhat flattened and are highly mobile are the ground beetles, some rove beetles, and others; their larvae are described as campodeiform. Some beetle larvae resemble hardened worms with dark head capsules and minute legs. These are elateriform larvae, and are found in the click beetle (Elateridae) and darkling beetle (Tenebrionidae) families. Some elateriform larvae of click beetles are known as wireworms. Beetles in the families of the Scarabaeoidea have short, thick larvae described as scarabaeiform, but more commonly known as grubs.

All beetle larvae go through several instars, which are the developmental stages between each moult. In many species the larvae simply increase in size with each successive instar as more food is consumed. In some cases, however, more dramatic changes occur. Among certain beetle families or genera, particularly those that exhibit parasitic lifestyles, the first instar (the planidium) is highly mobile in order to search out a host, while the following instars are more sedentary and remain on or within their host. This is known as hypermetamorphosis; examples include the blister beetles (family Meloidae) and some rove beetles, particularly those of the genus *Aleochara*.

Pupa

As with all endopterygotes, beetle larvae pupate, and from this pupa emerges a fully formed, sexually mature adult beetle, or imago. Adults have an extremely variable lifespan, from weeks to years, depending on the species. In some species the pupa may go through all four forms during its development, called hypermetamorphosis (for example,, Meloidae). Pupae always have no mandibles, or adecticous. In most, the appendages are not attached to the pupae, or they are exarate; with most being obtect in form.[5]

Behavior

Locomotion

Aquatic beetles use several techniques for retaining air beneath the water's surface. Beetles of the family Dytiscidae hold air between the abdomen and the elytra when diving. Hydrophilidae have hairs on their under surface that retain a layer of air against their bodies. Adult crawling water beetles use both their elytra and their hind coxae (the basal segment of the back legs) in air retention,[12] while whirligig beetles simply carry an air bubble down with them whenever they dive.

Photinus pyralis, firefly, in midflight.

The elytra allows beetles and weevils to both fly and move through confined spaces. Doings so by folding the delicate wings under the elytra while not flying, and folding their wings out just before take off. The unfolding and folding of the wings is operated by muscles attached to the wing base; as long as the tension on the radial and cubital veins remains, the wings remain straight. In day-flying species (for example, Buprestidae, Scarabaeidae), flight does not include large amounts of lifting of the elytra, having the metathorac wings extended under the lateral elytra margins.[1]

Communication

Beetles have a variety of ways to communicate. Some of which include a sophisticated chemical language through the use of pheromones. From the host tree, the mountain pine beetle have many forms of communication. They can emit both an aggregative pheromone and an anti-aggregative pheramone. The aggregative pheromone attracts other beetles to the tree, and the anti-aggregative pheromone neutralizes the aggregative pheromone. This helps to avoid the harmful effects of having too many beetles on one tree competing for resources. The mountain pine beetle can also stridulate to communicate, or rub body parts together to create sound, having a “scraper” on their abdomen that they rub against a grooved surface on the underside of their left wing cover to create a sound that is not audible to humans. Once the female beetles have arrived on a suitable pine tree host, they begin to stridulate and produce aggregative pheromones to attract other unmated males and females. New females arrive and do the same as they land and bore into the tree. As the males arrive, they enter the galleries that the females have tunneled, and begin to stridulate to let the females know they have arrived, and to also warn others that the female in that gallery is taken. At this point, the female stops producing aggregative pheromones and starts producing anti-aggregative pheromone to deter more beetles from coming.[13]

Since species of Coleoptera use environmental stimuli to communicate, they are affected by the climate. Microclimates, such as wind or temperature, can disturb the use of pheromones; wind would blow the pheromones while they ravel through the air. Stridulating can be interrupted when the stimulus is vibrated by something else.[13]

Parental care

Among insect, parental care is very uncommon, only found in a few species. Some beetles also display this unique social behavior.[1] One theory states why there is parental care is that it is necessary for the survival of the larvae, protecting them from adverse environmental conditions and predators. One species, a rover beetle (*Bledius spectabilis*) displays both causes for parental care: physical and biotic environmental factors. Said species lives in salt marshes, so the eggs and/or larvae are endangered by the rising tide. The maternal beetle will patrol the eggs and larva and apply the appropriate burrowing behavior the keep them from flooding and from asphyxiating. Another advantage is that the mother protects the eggs and larvae from the predatory carabid beetles species *Dicheirotrichus gustavi* and from the parasitoid wasp species *Barycnemis blediator*. Up to 15% of larvae are killed by this parasitoid wasp, being only protected by maternal beetles in their dens.[14]

a dung beetle rolling dung, near the giant tomb *Sa Ena 'e Thomes*, Sardinia, Italy

Some species of Dung beetles also display a form of parental care. Dung beetles, from which their name is derived, collect the feces, or "dung" and roll it into a ball, sometimes being up to 50 times their own weight; albeit sometimes it is also used to store food. Usually it is the male that rolls the ball, with the female hitch-hiking or simply following behind. In some cases the male and the female roll together. When a spot with soft soil is found, they stop and bury the dung ball. They will then mate underground. After the mating, both or one of them will prepare the brooding ball. When the ball is finished, the female lays eggs inside it, a form of mass provisioning. Some species do not leave after this stage, but remain to safeguard their offspring.[15]

Feeding

Mylabris pustulata (Meloidae) feeding on the petals of *Ipomoea carnea*

Besides being abundant and varied, beetles are able to exploit the wide diversity of food sources available in their many habitats. Some are omnivores, eating both plants and animals. Other beetles are highly specialized in their diet. Many species of leaf beetles, longhorn beetles, and weevils are very host-specific, feeding on only a single species of plant. Ground beetles and rove beetles (family Staphylinidae), among others, are primarily carnivorous and will catch and consume many other arthropods and small prey, such as earthworms and snails. While most predatory beetles are generalists, a few species have more specific prey requirements or preferences.[16]

Decaying organic matter is a primary diet for many species. This can range from dung, which is consumed by coprophagous species (such as certain scarab beetles of the family Scarabaeidae), to dead animals, which are eaten by necrophagous species (such as the carrion beetles of the family Silphidae). Some of the beetles found within dung and carrion are in fact predatory. These include the clown beetles, preying on the larvae of coprophagous and necrophagous insects.

Ecology

Defense and predation

Beetles may be preyed upon by other insects such as robber flies

Beetles and their larvae have a variety of strategies to avoid being attacked by predators or parasitoids. These include camouflage, mimicry, toxicity, and active defense. Camouflage involves the use of coloration or shape to blend into the surrounding environment. This sort of protective coloration is common and widespread among beetle families, especially those that feed on wood or vegetation, such as many of the leaf beetles (family Chrysomelidae) or weevils. In some of these species, sculpturing or various colored scales or hairs cause the beetle to resemble bird dung or other inedible objects. Many of those that live in sandy environments blend in with the coloration of the substrate.[10] :126 For example, the Giant African longhorn beetle (*Petrognatha gigas*) which resembles the moss and bark of the tree from which it feeds on.

Another defense that often uses color or shape to deceive potential enemies is mimicry. A number of longhorn beetles (family Cerambycidae) bear a striking resemblance to wasps, which helps them avoid predation even though the beetles are in fact harmless. This defense can be found to a lesser extent in other beetle families, such as the scarab beetles. Beetles may combine their color mimicry with behavioral mimicry, acting like the wasps they already closely resemble. Many beetle species, including ladybirds, blister beetles, and lycid beetles can secrete distasteful or toxic substances to make them unpalatable or even poisonous. These same species often exhibit aposematism, where bright or contrasting color patterns warn away potential predators, and there are, not surprisingly, a great many beetles and other insects that mimic these chemically protected species.[10]

Clytus arietis (Cerambycidae), a wasp mimic

Chemical defense is another important defense found amongst species of Coleoptera, usually being advertised by bright colors. Others may utilize behaviors that would be done when releasing noxious chemicals (for example,, Tenebrionidae). Chemical defense may serve purposes other than just protection from vertebrates, such as protection from a wide range of microbes, and repellents. Some species release chemicals in the form of a spray with surprising accuracy, such as ground beetles (Carabidae), may spray chemicals from their abdomen to repel predators. Some species take advantage of the plants from which they feed, and sequester the chemicals from the plant that would protect it and incorporate into their own defense. African carabid beetles (for example,, *Anthia* and *Thermophilium*) employ the same chemicals used by ants, while Bombardier beetles have a their own unique separate gland, spraying potential predators from far distances.[10] :126

Large ground beetles and longhorn beetles may defend themselves using strong mandibles and/or spines or horns to forcibly persuade a predator to seek out easier prey.[10] :126 Many species have large protrusions from their thorax and head such as the Rhinoceros beetle, which can be used to defended themselves from predators. Many species of weevil that feed out in the open on leaves of plants react to attack by employing a "drop-off reflex". Even further, some will combine it with thanatosis, which they will close up their legs, antennae, mandibles, ect. and use their cryptic coloration to blend in with the background. Species with varied coloration do not do this as they can not camaflouge.[1] :199

Parasitism

There are over 1000 known species of beetles to be either parasitic, predatory, or commensals in the nests of ants.[17] Most beetle larvae can be considered parasites, as they feed on the plants and live inside the bark of trees and plants. Though there are a few species that are ectoparasitic to mammals, such as *Platypsyllus castoris*, which is affects beavers (**Castor** spp.). This beaver beetle parasite is a parasite as both an adult and a larva, with the appropriate modifications. They are wingless and eyeless with a striking dorso-ventral flattening. Additionally, P. castoris, has a remarkably modified antennal club, with antennomeres 3-11 shortened, globularly compacted, and partly enclosed in a scoop shaped antennomere 2, as is also found in **Gyrinidae** and **Dryopidae**.[18]

other parasites include those who are parasitoids of other invertebrates, such as the small hive beetle (*Aethina tumida*) infecting Honey bee hives. The larvae tunnel through comb towards stored honey or pollen, damaging or destroying cappings and comb in the process. Larvae defecate in honey and the honey becomes discolored from the feces, which causes fermentation and a frothiness in the honey; the honey develops a characteristic odor of decaying oranges. Damage and fermentation cause honey to run out of combs, destroing large portions in hives and sometimes the extracting rooms. Heavy infestations cause bees to abscond; some beekeepers have reported the rapid collapse of even strong colonies.[19]

Pollination

Beetle-pollinated flowers are usually large, greenish or off-white in color and heavily scented. Scents may be spicy, fruity, or similar to decaying organic material. Most beetle-pollinated flowers are flattened or dish shaped, with pollen easily accessible, although they may include traps to keep the beetle longer. The plant's ovaries are usually well protected from the biting mouthparts of their pollinators.[7] Beetles may be particularly important in some parts of the world such as semi-arid areas of southern Africa and southern California[20] and the montane grasslands of KwaZulu-Natal in South Africa.[21]

Mutualism

Amongst most orders of insects, mutualism is not common, however there are some examples in species of Coleoptera. Such as the Ambrosia beetle, the Ambrosia fungus, and probably bacteria. The beetles excavate tunnels in dead trees in which they cultivate fungal gardens, their sole source of nutrition. After landing on a suitable tree, an ambrosia beetle excavates a tunnel in which it releases spores of its fungal symbiont. The fungus penetrates the plant's xylem tissue, digests it, and concentrates the

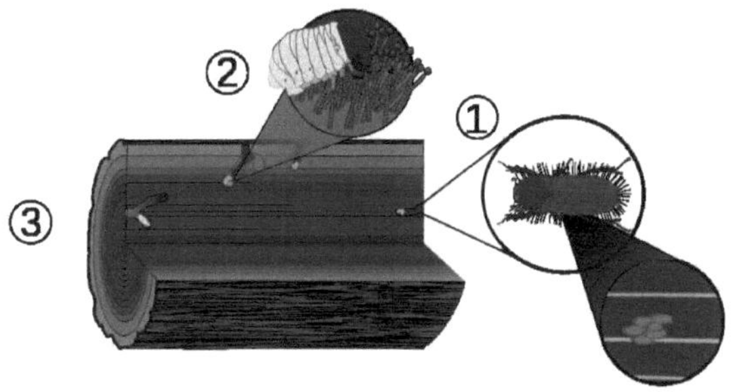

1: The adult beetle burrows hole into wood and lays eggs, carrying fungal spores in their mycangia
2: The larva feeds on the fungus which digest the wood, removing toxins: they mutually benefit
3: The larva pupates and then ecloses.

nutrients on and near the surface of the beetle gallery; so the weevils and the fungus both benefit. The beetles can not eat due to toxins, which uses its relationship with fungi to help overcome it's host tree defenses and to provide nutrition for their larvae.[22]

The beetle-fungal mutualism is chemically mediated by a bacterially produced polyunsaturated peroxide. The molecule's selective toxicity toward the beetle's fungal antagonist, combined with the prevalence and localization of its bacterial source, indicates an insect-microbe association that is both mutualistic and coevolved. This unexpected finding in a well-studied system indicates that mutualistic associations between insects and antibiotic-producing bacteria are more common than currently recognized and that identifying their small-molecule mediators can provide a powerful search strategy for therapeutically useful antimicrobial compounds.[22] [23]

Commensalism

Pseudoscorpions are small arachnids with a flat, pear-shaped body and pincers that resemble those of scorpions (only distant relatives), usually ranging from 2 to 8 millimetres (0.08 to 0.31 in) in length.[24] Their small size allows them to hitch rides under the elytra of a giant harlequin beetle to be dispersed over wide areas while simultaneously being protected from predators. They may also find mating partners as other individuals join them on the beetle. This would be a form of parasitism if the beetle was harmed in the process, however the beetle is, presumably, unaffected by the presence of the hitchhikers.[25] [26]

Phylogeny and systematics

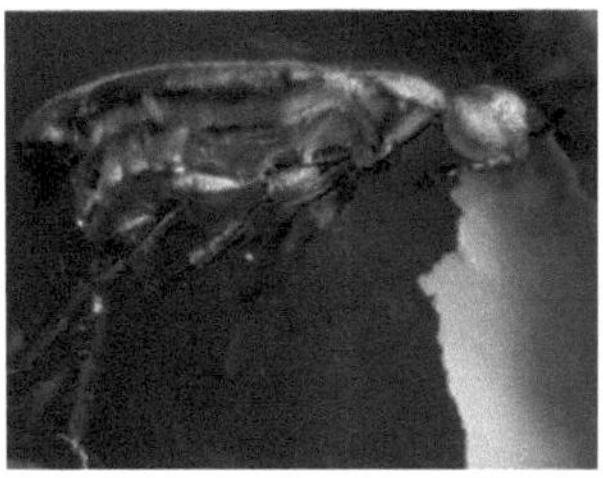

Baltic amber inclusions, from the Eucene era, 50 million years old (Coleoptera, Scraptiidae)

Fossil record

A 2007 study based on DNA of living beetles and maps of likely beetle evolution indicated that beetles may have originated during the Lower Permian, up to 299 million years ago.[27] In 2009, a fossil beetle was described from the Pennsylvanian of Mazon Creek, Illinois, pushing the origin of the beetles to an earlier date, 318 to 299 [28] million years ago.[28] Fossils from this time have been found in Asia and Europe, for instance in the red slate fossil beds of Niedermoschel near Mainz, Germany.[29] Further fossils have been found in Obora, Czechia and Tshekarda in the Ural mountains, Russia.[30] However, there are only a few fossils from North America before the middle Permian, although both Asia and North America had been united to Euramerica. The first discoveries from North America were made in the Wellington formation of Oklahoma and were published in 2005 and 2008.[31] [32]

As a consequence of the P-Tr Mass Extinction at the border of Permian and Triassic, there is only little fossil record of insects including beetles from the Lower Triassic.[33] However, there are a few exemptions, like in Eastern Europe: At the Babiy Kamen site in the Kuznetsk Basin numerous beetle fossils were discovered, even entire specimen of the infraorders Archostemata (e.i., **Ademosynidae**, **Schizocoleidae**), Adephaga (e.i., **Triaplidae**, **Trachypachidae**) and Polyphaga (e.i., **Hydrophilidae**, **Byrrhidae**, **Elateroidea**) and in nearly a perfectly preserved condition.[34] However, species from the families **Cupedidae** and **Schizophoroidae** are not present at this site, whereas they dominate at other fossil sites from the Lower Triassic. Further records are known from Khey-Yaga, Russia in the Korotaikha Basin.[31] There are many important sites from the Jurassic, with more than 150 important sites with beetle fossils, the majority being situated in Eastern Europe and North Asia. In North America and especially in South America and Africa the number of sites from that time period is smaller and the sites have not been exhaustively investigated yet. Outstanding fossil sites include Solnhofen in Upper Bavaria, Germany,[35] Karatau in South Kazakhstan,[36] the Yixian formation in Liaoning, North China[37] as well as the

Jiulongshan formation and further fossil sites in Mongolia. In North America there are only a few sites with fossil records of insects from the Jurassic, namely the shell limestone deposits in the Hartford basin, the Deerfield basin and the Newark basin.[31] [38]

There is a large number of important fossil sites worldwide containing beetles from the Cretaceous. Most of them are located in Europe and Asia and belong to the temperate climate zone during the Cretaceous. A few of the fossil sites mentioned in the chapter Jurassic also shed some light on the early cretaceous beetle fauna (for example, the Yixian formation in Liaoning, North China).[37] Further important sites from the Lower Cretaceous include the Crato Fossil Beds in the Araripe basin in the Ceará, North Brazil as well as overlying Santana formation, with the latter was situated near the paleoequator, or the position of the earth's equator in the geologic past as defined for a specific geologic period. In Spain there are important sites near Montsec and Las Hoyas. In Australia the Koonwarra fossil beds of the Korumburra group, South Gippsland, Victoria is noteworthy. Important fossil sites from the Upper Cretaceous are Kzyl-Dzhar in South Kazakhstan and Arkagala in Russia.[31]

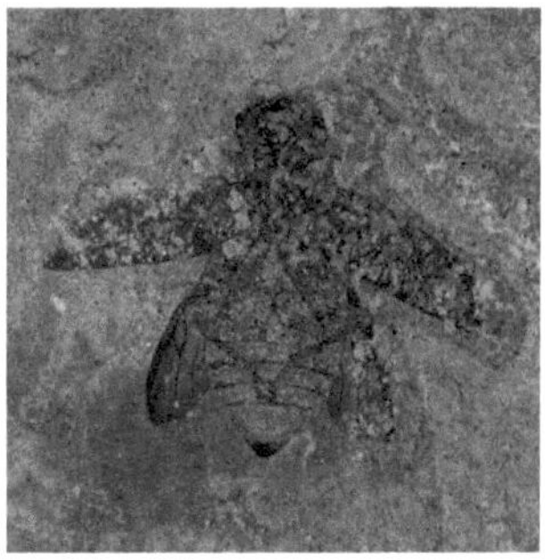

Fossil buprestid beetle from the Eocene Messel pit, which retains its structural color

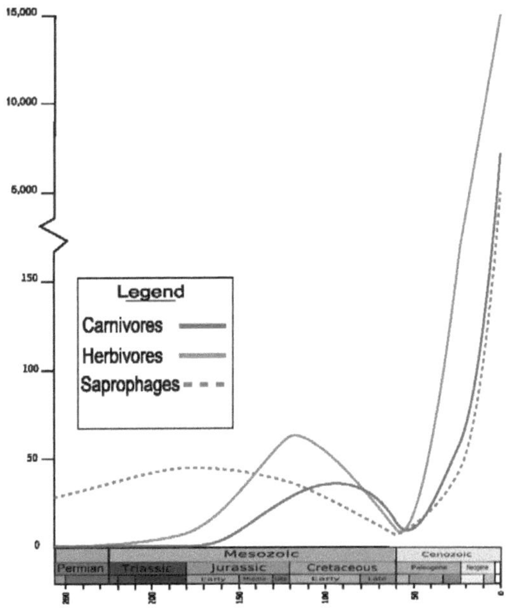

The phylogenetic growth of three different trophic levels in Coleoptera by number of genera.

Evolution

The oldest known insect that resembles species of Coleoptera date back to the Lower Permian (270 mya), though they instead have 13-segmented antennae, elytra with more fully developed venation and more irregular longitudinal ribbing, and an abdomen and ovipositor extending beyond the apex of the elytra. The oldest true beetle, that is having features that include 11-segmented antennae, regular longitudinal ribbing on the elytra, and having genitalia that are internal.[1] :186 At the end of the Permian, the biggest mass extinction in the history history took place, collectively called the Permian–Triassic extinction event: 30% of all insect species became extinct, however, it is the only mass extinction of insects in Earth's history until today.[31]

Consequence do to the P-Tr extinction, there is only little fossil record of insects including beetles from the Lower Triassic (220 mya). Around this time, during the Late Triassic, mycetophagous, or fungus feeding species (e.i. **Cupedidae**) appear in the fossil record. In the stages of the Upper Triassic representatives of the algophagous, or algae feeding species (e.i. **Triaplidae** and **Hydrophilidae**) begin to appear, as well as predatory water beetles. The first primitive weevils appear (e.i. **Obrienidae**), as well as the first representatives of the rove beetles (e.i. **Staphylinidae**), which show no marked difference in physique compared to recent species.[31]

During the Jurassic (210-145 mya) there was a dramatic increase in the known diversity of family-level Coleoptera.[31] This includes the development and growth of carnivorous and herbivorous species. Species of the superfamily **Chrysomeloidea** are believed to have developed around the same time, which include a wide array of plant host ranging from cycads and conifers, to angiosperms.[1] :186 Close to the Upper Jurassic, the portion of the **Cupedidae** decreased, however at the same time the diversity of the early plant eating, or phytophagous species increased. Most of the recent phytophagous species of Coleoptera feed on flowering plants or angiosperms. It is believed that the increase in diversity of the angiosperms also influenced the diversity of the phytophagous species, which doubled during the Middle Jurassic. However, recently doubts have been raised since the increase of the number of beetle families during the Cretaceous does not correlate with the increase of the number of angiosperm species.[39] Also around the same time, numerous primitive weevils (e.i., **Curculionoidea**) and click beetles (e.i. *Elateroidea*) appeared. Also first jewel beetles (e.i., **Buprestidae**) are present, however, they were rather rare until the Cretaceous.[40] [41] [42] The first scarab beetles would appear around this time, however they were not coprophagous, or feeding upon fecal matter, presumably feeding upon the rotting wood with the help of fungus, and early example of a mutualistic relationship (see the Mutualism section).

The Cretaceous witness the initiation of the most recent round of southern landmass fragmentation, via the opening of the southern Atlantic ocean and the isolation of New Zealand, while the South America, Antarctica, and Australia grew more distant.[1] :186 During the Cretaceous the diversity of **Cupedidae** and **Archostemata** decreased considerably. Predatory ground beetles (**Carabidae**) and rove beetles (**Staphylinidae**) began to distribute into different patterns: whereas the **Carabidae** predominantly occurred in the warm regions, the **Staphylinidae** and click beetles (**Elateridae**) preferred many areas with temperate climate. Likewise, predatory species of **Cleroidea** and **Cucujoidea**, hunted their prey under the bark of trees together with the jewel beetles (**Buprestidae**). The jewel beetles diversity increased rapidly during the Cretaceous, as they were the primary consumers of wood,[43] while longhorn beetles (**Cerambycidae**) were rather rare and their diversity increased only towards the end of the Upper Cretaceous.[31] The first coprophagous beetles have been recorded from the Upper Cretaceous,[44] and are believed to have lived on the excrement of herbivorous dinosaurs, however there is still a discussion, whether the beetles were always tied to mammals during its development.[45] Also, the first species with an adaption of both larvae and adults to the aquatic lifestyle are found. Whirligig beetles (**Gyrinidae**) were moderately diverse, although other early beetles (e.i., **Dytiscidae**) were less, with the most widespread being the species of **Coptoclavidae**, which preyed on aquatic fly larvae.[31]

The time between the Paleogene and the Neogene, or more recent history is where today's beetles developed. During this time, the continents began to situate themselves to where we see them today. Around 5 million years ago the land bridge between South America and North America was formed, and this is when fauna exchange between Asia and North America started. Even though many recent genera and species already existed during the Miocene, however, their distribution differed considerably from today's.[31]

Phylogeny

The suborders diverged in the Permian and Triassic. Their phylogenetic relationship is uncertain, with the most popular hypothesis being that Polyphaga and Myxophaga are most closely related, with Adephaga as the sister group to those two, and Archostemata as sister to the other three collectively.[1] :186[46] Although there are six other competing hypotheses, the other most widely discussed one is Myxophaga as the sister group of all remaining beetles rather than just of Polyphaga.[47] Evidence for a close relationship of the two suborders, Polyphaga and Myxophaga, includes the shared reduction in the number of larval leg articles. The Adephaga is further considered as sister to Myxophaga and Polyphaga, based on their completely sclerotized elytra, reduced number of crossveins in the hind wings, and the folded (as opposed to rolled) hind wings of those three suborders.

Recent cladistic analysis of some of the structural characteristics supports the Polyphaga and Myxophaga hypothesis.[46] The membership of the clade Coleoptera is not in dispute, with the exception of the twisted-wing parasites, Strepsiptera. These odd insects have been regarded as related to the beetle families Rhipiphoridae and Meloidae, with which they share first-instar larvae that are active, host-seeking triungulins and later-instar larvae that are endoparasites of other insects, or the sister group of beetles, or more distantly related to insects.[47] [48]

Taxonomy

There are about 450,000 species of beetles – representing about 40% of all known insects. Such a large number of species poses special problems for classification, with some families consisting of thousands of species and needing further division into subfamilies and tribes. This immense number of species allegedly led evolutionary biologist J. B. S. Haldane to quip, when some theologians asked him what could be inferred about the mind of the Creator from the works of His Creation, that God displayed "an inordinate fondness for beetles".[49]

- Polyphaga is the largest suborder, containing more than 300,000 described species in more than 170 families, including rove beetles (Staphylinidae), scarab beetles (Scarabaeidae), blister beetles (Meloidae), stag beetles (Lucanidae) and true weevils (Curculionidae). These beetles can be identified by the cervical sclerites (hardened parts of the head used as points of attachment for muscles) absent in the other suborders.
- Adephaga contains about 10 families of largely predatory beetles, includes ground beetles (Carabidae), Dytiscidae and whirligig beetles (Gyrinidae). In these beetles, the testes are tubular and the first abdominal sternum (a plate of the exoskeleton) is divided by the hind coxae (the basal joints of the beetle's legs).
- Archostemata contains four families of mainly wood-eating beetles, including reticulated beetles (Cupedidae) and the telephone-pole beetle.
- Myxophaga contains about 100 described species in four families, mostly very small, including Hydroscaphidae and the genus *Sphaerius*.

Relationship to people

As pests

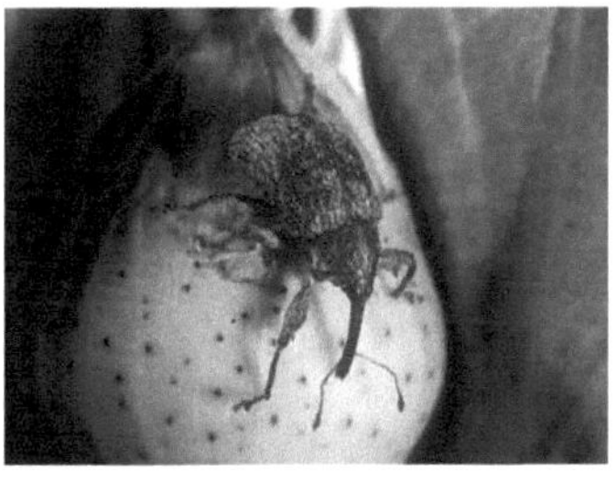

About $^{3}/_{4}$ of beetle species are phytophagous in both the larval and adult stages, living in or on plants, wood, fungi, and a variety of stored products, including cereals, tobacco, and dried fruits. Because many of these plants are important for agriculture, forestry, and the household, the beetle can be considered a pest.[5] Some of these species cause significant damage, such as the Boll weevil, which feeds on cotton buds and flowers. The Boll Weevil crossed the Rio Grande near Brownsville, Texas to enter the United States from Mexico around 1892[50] and had reached southeastern Alabama by 1915. By the mid 1920s it had entered all cotton growing regions in the U.S., traveling 40 to 160 miles (60–260 km) per year. It remains the most destructive cotton pest in North America. Mississippi State University has estimated that since the boll weevil entered the United States it has cost U.S. cotton producers about $13 billion, and in recent times about $300 million per year.[50] Many other species also have done extensive damage to plant populations, such as the Bark beetle and Elm Leaf beetle. The Bark beetle and Elm Leaf beetle, among other species, have been known to nest in Elm trees. Bark beetles in particular carry Dutch elm disease as they move from infected breeding sites to feed on healthy elm trees. The spread of Dutch elm disease by the beetle has led to the devastation of elm trees in many parts of the Northern Hemisphere, notably in Europe and North America.[51]

Situations in which a species has developed immunity to pesticides are worse. Such as the Colorado potato beetle, *Leptinotarsa decemlineata*, which is a notorious pest of potato plants. Crops are destroyed and the beetle can only be treated by employing expensive pesticides, many of which it has begun to develop resistance to. As well as potatoes, suitable hosts can be a number of plants from the potato family (Solanaceae), such as nightshade, tomato, aubergine and capsicum. The Colorado potato beetle has developed resistance to all major insecticide classes, although not every population is resistant to every chemical.[52]

Larvae of the Colorado potato beetle, *Leptinotarsa decemlineata*

Pest don't only affect agriculture, but can also even affect houses, such as the Death watch beetle. The death watch beetle, *Xestobium rufovillosum*, (family Anobiidae) is of considerable importance as a pest of older wooden buildings in Great Britain. It attacks hardwoods such as oak and chestnut, always where some fungal decay has taken or is taking place. It is thought that the actual introduction of the pest into buildings takes place at the time of construction.[53]

Other pest include the Coconut hispine beetle, *Brontispa longissima*, feeds on young leaves and damages seedlings and mature coconut palms. On September 27, 2007, Philippines' Metro Manila and 26 provinces were quarantined due to having been infested with this pest (to save the $800-million Philippine coconut industry).[54] The mountain pine beetle normally attacks mature or weakened lodgepole pine. It can be the most destructive insect pest of mature pine forests. The current infestation in British Columbia is the largest Canada has ever seen.[55]

As beneficial

Coccinella septempunctata, a beneficial beetle

Beetles are not only pests, but can also be beneficial, usually by controlling the populations of pests. One of the best, and widely known, examples are the Ladybugs or ladybirds (family Coccinellidae). Both the larvae and adults are found feeding on aphid colonies. Other ladybugs feed on scale insects and mealybugs. If normal food sources are scarce, they may feed on other things, such as small caterpillars, young plant bugs, honeydew and nectar.[56] Ground beetles (family Carabidae) are common predators of many different insects and other arthropods, including fly eggs, caterpillars, wireworms and others.[57]

Dung beetles (Coleoptera, Scarabidae) have been successfully used to reduce the populations of pestilent flies and parasitic worms that breed in cattle dung. The beetles make the dung unavailable to breeding pests by quickly rolling and burying it in the soil, with the added effect of improving soil fertility and nutrient cycling. The Australian Dung Beetle Project (1965–1985), led by Dr. George Bornemissza of the Commonwealth Scientific and Industrial Research Organization introduced species of dung beetle to Australia from South Africa and Europe and effectively reduced the bush fly (*Musca vetustissima*) population by 90%.

Dung beetles play a remarkable role in agriculture. By burying and consuming dung, they improve nutrient recycling and soil[58] structure. They also protect livestock, such as cattle, by removing the dung which, if left, could provide habitat for pests such as flies. Therefore, many countries have introduced the creature for the benefit of animal husbandry. In developing countries, the beetle is especially important as an adjunct for improving standards of hygiene. The American Institute of Biological Sciences reports that dung beetles save the United States cattle industry an estimated US$380 million annually through burying above-ground livestock faeces.[59]

Some beetles help in a professional setting, doing things that people can't; such as those of the family Dermestidae are often used in taxidermy and preparation of scientific specimens to clean bones of remaining soft tissue. The beetle larvae to clean skulls because they do a thorough job of cleaning, and the beetle larvae do not leave the tool marks that taxidermists tools do. Another benefit is that with no traces of meat remaining, and no emulsified fats in the bones, the trophy will not develop the unpleasant dead order. Using the beetle larvae means that all cartilage is removed along with the flesh, leaving the bones spotless.[60]

As food

Mealworms presented in a bowl for human consumption.

Insects are used as human food in 80% of the world's nations.[61] Beetles are the most widely eaten insects. 344 species are known to be used as food. They are usually eaten in the larval stage.[62] :5 The mealworm is the most eaten beetle species. The larvae of the darkling beetle and the rhinoceros beetle are also commonly eaten.

In art

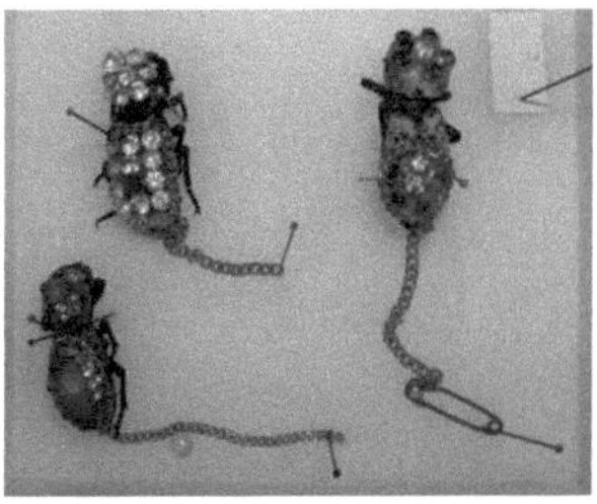

Zopheridae examples of jewelry taken at the Texas A&M University Insect Collection in College Station, Texas.

Many beetles have beautiful and durable elytra that have been used as a material in arts, with Beetlewing the best example.[63] Sometimes they're also incorporated into ritual objects for their religious significance. Whole beetles, either by themselves or encased in clear plastic, are also made into everything from cheap souvenirs such as key chains to expensive fine-art jewelry. In parts of Mexico, beetles of the genus Zopherus are made into living brooches by attaching costume jewelry and golden chains. This is made possible by the incredibly hard elytra and sedentary habits of the genus.[64]

In ancient culture

ḫpr in hieroglyphs

Beetles are prominent in ancient culture, however the most would probably be the Dung beetle in Ancient Egyptian. Several species of the dung beetle, most notably the species *Scarabaeus sacer* (often referred to as *the sacred scarab*), enjoyed a sacred status among the ancient Egyptians.[65] Popular interpretation in modern academia theorizes the hieroglyphic image of the beetle represents a triliteral phonetic that Egyptologists transliterate as *xpr* or ḫpr and translate as "to come into being", "to become" or "to transform". The derivative term *xprw* or ḫpr(w) is variously translated as "form", "transformation", "happening", "mode of being" or "what has come into being", depending on the context. It may have existential, fictional, or ontologic significance.[66]

A scarab statue at the Karnak temple complex

The scarab was linked to Khepri ("he who has come into being"), the god of the rising sun. The ancients believed that the dung beetle was only male in gender, and reproduced by depositing semen into a dung ball. The supposed self-creation of the beetle resembles that of Khepri, who creates himself out of nothing. Moreover, the dung ball rolled by a dung beetle resembles the sun. Plutarch wrote:

A scarab, depicted on the walls of Tomb KV6 in the Valley of the Kings

> "The race of beetles has no female, but all the males eject their sperm into a round pellet of material which they roll up by pushing it from the opposite side, just as the sun seems to turn the heavens in the direction opposite to its own course, which is from west to east."[67]

The ancient Egyptians believed that Khepri renewed the sun every day before rolling it above the horizon, then carried it through the other world after sunset, only to renew it, again, the next day. Some New Kingdom royal tombs exhibit a threefold image of the sun god, with the beetle as symbol of the morning sun. The astronomical ceiling in the tomb of Ramses VI portrays the nightly "death" and "rebirth" of the sun as being swallowed by Nut, goddess of the sky, and re-emerging from her womb as Khepri.[65]

Excavations of ancient Egyptian sites have yielded images of the scarab in bone, ivory, stone, Egyptian faience, and precious metals, dating from the Sixth Dynasty and up to the period of Roman rule. They are generally small, bored to allow stringing on a necklace, and the base bears a brief inscription or cartouche. Some have been used as seals. Pharaohs sometimes commissioned the manufacture of larger images with lengthy inscriptions, such as the commemorative scarab of Queen Tiye. Massive sculptures of scarabs can be seen at Luxor Temple, at the Serapeum in Alexandria (see Serapis) and elsewhere in Egypt.

The scarab was of prime significance in the funerary cult of ancient Egypt.[68] Scarabs, generally, though not always, were cut from green stone, and placed on the chest of the deceased. Perhaps the most famous example of such "heart scarabs" is the yellow-green pectoral scarab found among the entombed provisions of Tutankhamen. It was carved from a large piece of Libyan desert glass. The purpose of the "heart scarab" was to ensure that the heart would not bear witness against the deceased at judgement in the Afterlife. Other possibilities are suggested by the "transformation spells" of the *Coffin Texts*, which affirm that the soul of the deceased may transform (*xpr*) into a human being, a god, or a bird and reappear in the world of the living.

In contrast to funerary contexts, some of ancient Egypt's neighbors adopted the scarab motif for seals of varying types. The best-known of these being Judean LMLK seals (8 of 21 designs contained scarab beetles), which were used exclusively to stamp impressions on storage jars during the reign of Hezekiah. The scarab remains an item of popular interest thanks to modern fascination with the art and beliefs of ancient Egypt. Scarab beads in semiprecious stones or glazed ceramics can be purchased at most bead shops, while at Luxor Temple a massive ancient scarab has been roped off to discourage visitors from rubbing the base of the statue "for luck".

In modern culture

Beetles still play a role in modern culture such as their use in insect fighting. Normally exploiting the large beetle's territorial behavior. Individuals may hoard many different type of insects for the fights. Some of the most popular are the Stag Beetle, Rhinoceros Beetle, Kabutomushi, and Goliath Beetle, as their sheer size and jumping ability make them formidable opponents. They are trained by their owners to become stronger and more aggressive.

The study of beetles is called **coleopterology** (from *Coleoptera*, see above, and Greek -λογία, *-logia*), and its practitioners are *coleopterists*. Coleopterists have formed organizations to facilitate the study of beetles. Among these is The Coleopterists Society, an international organization based in the United States. Such organizations may have both professionals and amateurs interested in beetles as members. Research in this field is often published in peer-reviewed journals specific to the field of coleopterology, though journals dealing with general entomology also publish many papers on various aspects of beetle biology. Some of the journals specific to beetle research are:

- *The Coleopterist* (United Kingdom beetle fauna)
- *The Coleopterists Bulletin* (published by The Coleopterists Society)
- *Elytron* (published by the European Association of Coleopterology)

See also

- Blister beetle dermatitis
- Heteroptera, suborder of the Hemiptera (true bugs), which may resemble beetles
- Leaf litter sieve

References

General references

- Poul Beckmann, *Living Jewels: The Natural Design of Beetles* ISBN 3-7913-2528-0
- Arthur V. Evans, Charles Bellamy, and Lisa Charles Watson, *An Inordinate Fondness for Beetles* ISBN 0-520-22323-3
- Cooter J. & Barclay M.V.L. (eds.) (2006) A Coleopterist's Handbook. Amateur Entomological Society. 439 pages. ISBN 0-900054-70-0
- Entomological Society of America, *Beetle Larvae of the World* ISBN 0-643-05506-1
- David Grimaldi, Michael S. Engel, *Evolution of the Insects* ISBN 0-521-82149-5
- Ross H. Arnett, Jr. and Michael C. Thomas, *American Beetles* (CRC Press, 2001–2002). ISBN 0-8493-1925-0
- K. W. Harde, *A Field Guide in Color to Beetles* ISBN 0-7064-1937-5 Pages 7–24
- White, R.E. 1983. Beetles. Houghton Mifflin Company, New York, NY. ISBN 0-395-91089-7

Cited references

[1] Powell, Jerry A. (2009). "Coleoptera" (http://books.google.co.in/books?id=wrMcPwAACAAJ). In Resh, Vincent H.; Cardé, Ring T.. *Encyclopedia of Insects* (2 (illustrated) ed.). Academic Press. pp. 1132. ISBN 9780123741448. . Retrieved 14 November 2010.

[2] P. M. Hammond 1992. Species inventory. pp. 17–39 in Global Biodiversity, Status of the Earth's Living Resources, B. Groombridge, ed. Chapman and Hall, London. 585 pp.

[3] Arthur D. Chapman (2009) (PDF). *Numbers of Living Species in Australia and the World* (http://www.environment.gov.au/biodiversity/abrs/publications/other/species-numbers/2009/pubs/nlsaw-2nd-complete.pdf) (2nd ed.). Department of the Environment, Water, Heritage and the Arts. ISBN 978-0-642-56861-8. .

[4] Harper, Douglas. "Coleoptera" (http://www.etymonline.com/index.php?term=Coleoptera). The Online Etymology Dictionary. . Retrieved 26 February 2011.

[5] Gilliott, Cedric (August 1995). *Entomology* (http://books.google.com/books?id=DrTKxvZq_IcC&pg=PA96&dq=Insect+classification+based+on+winged+and+wingless#v=onepage&q=coleoptera&f=false) (2 ed.). Springer-Verlag New York, LLC. ISBN 0-306-44967-6. .

[6] Foottit, Robert G.; Peter Holdridge Adler (2009). *Insect biodiversity: science and society* (http://books.google.com/books?id=LBZHpYY2_8gC&printsec=frontcover#v=snippet&q=Biodiversity of Coleoptera&f=false). John Wiley and Sons. ISBN 1405151420. .

[7] Gullan, P.J.; P.S. Cranston (March 22, 2010). *The Insects: An Outline of Entomology* (http://books.google.com/?id=S7yGZasJ7nEC&printsec=frontcover&dq=Insects&cd=1#v=onepage&q=&f=false) (4 ed.). Oxford: Wiley, John & Sons, Incorporated. ISBN 1-444-33036-5. .

[8] Michael A. Ivie (2002). Ross H. Arnett & Michael Charles Thomas. ed. *American Beetles: Polyphaga: Scarabaeoidea through Curculionoidea*. Volume 2 of American Beetles. CRC Press. ISBN 9780849309540.

[9] Schmidt-Nielsen, Knut (Jan 15, 1997). "Insect Respiration" (http://books.google.com/?id=Af7IwQWJoCMC&pg=PA55&lpg=PA55&dq=oxygen+diffusing+into+a+bubble#v=onepage&q&f=false). *Animal Physiology: Adaptation and Environment* (5 (illustrated) ed.). Cambridge University Press. pp. 55. ISBN 0521570980. . Retrieved 6 March 2010.

[10] Evans, Arthur V.; Bellamy, Charles (April 2000). *An Inordinate Fondness for Beetles* (http://books.google.com/?id=ZZ_hfpMo8oAC&pg=PA31&dq=The+Beetles+of+the+World#v=onepage&q=defense&f=false). ISBN 0-520-22323-3. .

[11] Scoble, MJ. (1992). *The Lepidoptera: Form, function, and diversity.*. Oxford Univ. Press. ISBN 9781402062421.

[12] R. H. Arnett, Jr. & M. C. Thomas (2001). "Haliplidae". *American Beetles, Volume 1*. CRC Press, Boca Raton, Florida. pp. 138–143. ISBN 0-8493-1925-0.

[13] "Mountain Pine Beetle - Beetle Love" (http://www.pc.gc.ca/eng/docs/v-g/dpp-mpb/sec2/dpp-mpb2b.aspx). Parks Canada. . Retrieved March 13, 2011.

[14] Brill, Leiden E.J. (Oct., 1989). "Observations on certain lepidopterous and hymenopterous parasites of Polistes wasps". *Behaviour* **110** (1/4): pp. 76–92 (1–4). JSTOR 4534785.

[15] Hanski, Ilkka; Yves, Cambefort (1991). *Dung Beetle Ecology*. Princeton University Press. pp. 626–672. ISBN 0691087393.

[16] A.L. Lobanov (2002). "feeding" (http://www.zin.ru/animalia/coleoptera/eng/biol3.htm). *Beetle Biology And Ecology*. Beetles (Coleoptera) and Coleopterologist. . Retrieved March 13, 2011.

[17] Meyer, John R. (8 March 2005). "Coleoptera" (http://www.cals.ncsu.edu/course/ent425/compendium/coleop~1.html). Department of Entomology, NC State University. . Retrieved March 13, 2011.

[18] Peck, Stewart B. (2006). "Distribution and biology of the ectoparasitic beaver beetle Platypsyllus castoris Ritsema in North America (Coleoptera: Leiodidae: Platypsyllinae)" (http://digitalcommons.unl.edu/cgi/viewcontent.cgi?article=1107&context=insectamundi). *Insecta Mundi* (Carleton University, Canada) **20** (1–2). . Retrieved March 13, 2011.

[19] Neumann P, Elzen PJ (2004). "The biology of the small hive beetle (*Aethina tumida*, Coleoptera: Nitidulidae): Gaps in our knowledge of an invasive species". *Apidologie* **35** (3): 229–247. doi:10.1051/apido:2004010.

[20] Jones, GD and SD Jones (2001). "The uses of pollen and its implication for Entomology". Neotropical Entomology 30 (3): 314–349. doi:10.1590/S1519-566X2001000300001

[21] Ollerton J., Johnson S. D., Cranmer, L. and Kellie, S. (2003) The pollination ecology of an assemblage of grassland asclepiads in South Africa. Annals of Botany 92: 807-834.

[22] Malloch, D., and M. Blackwell. 1993. Dispersal biology of ophiostomatoid fungi. p. 195-206. In: Ceratocystis and Ophiostoma: Taxonomy, Ecology and Pathology. Eds., Wingfield, M.J., K.A. Seifert, and J.F. Webber. APS, St. Paul.

[23] Francke-Grossmann H. (1967). *Ectosymbiosis in wood inhabiting insects*. In: M. Henry (ed.) *Symbiosis*, Vol. 2. Academic Press, NewYork. pp.141-205

[24] "Pseudoscorpions" (http://www.ento.psu.edu/extension/factsheets/pseudoscorpion.htm). *Insect Advice from Extension*. College of Agricultural Sciences. 2011. . Retrieved March 13, 2011.

[25] Gail, Vines (18 April 1992). "Hitchhiking pseudoscorpions take beetles for a ride" (http://www.newscientist.com/article/mg13418173.500--science-hitchhiking-pseudoscorpions-take-beetles-for-a-ride-.html). *New Scientist* (1817). .

[26] Poole, Robert W. (2004). "Ecology - Population Ecology - Commensalism" (http://www.nearctica.com/ecology/pops/commens.htm#pseudoscorpion). *Nearctica*. .

[27] Dave Mosher (December 26, 2007). "Modern beetles predate dinosaurs" (http://www.livescience.com/animals/071226-tough-beetles.html). Live Science. . Retrieved June 24, 2010.

[28] Oliver Béthoux (2009). "The earliest beetle identified". *Journal of Paleontology* **83** (6): 931–937. doi:10.1666/08-158.1.

[29] Hörnschemeyer, T.; H. Stapf, Terra Nostra. "Die Insektentaphozönose von Niedermoschel (Asselian, unt. Perm; Deutschland)" (in German). *Schriften der Alfred-Wegener-Stiftung* (99/8): 98.

[30] Moravia, J; Kukalová, Sb. Geol. Ved. Rada. P. (1969). "On the systematic position of the supposed Permian beetles, Tshecardocoleidae [sic], with a description of a new collection". *Paleontol* (11): 139–161.

[31] Benisch, Christoph (2010). "Phylogeny of the beetles" (http://www.kerbtier.de/Pages/Themenseiten/enPhylogenie.html#L9). *The beetle fauna of Germany*. Kerbtier. . Retrieved March 16, 2011.

[32] Beckemeyer, R. J.; M. S. Engel (2008). "A Second Specimen of Permocoleus (Coleoptera) from the Lower Permian Wellington Formation of Noble County, Oklahoma" (http://fossilinsects.net/pdfs/beckemeyer_engel_2008_JKansEntSoc_PermocoleusPermOklahoma.pdf). *Journal of the Kansas Entomological Society* **81** (1): 4–7. doi:10.2317/JKES-708.01.1. . Retrieved 2011-03-17.

[33] Shcherbakov, D. E. (2008). "On Permian and Triassic Insect Faunas in Relation to Biogeography and the Permian-Triassic Crisis". *Paleontological Journal* **42** (1): 15–31..

[34] Ponomarenko, A. G. (2004). "Beetles (Insecta, Coleoptera) of the Late Permian and Early Triassic" (http://palaeoentomolog.ru/Publ/PALS185.pdf). *Paleontological Journal* **38** (Suppl. 2): S185–S196. . Retrieved 2011-03-17.

[35] Vienna, A. G (1985). "Fossil insects from the Tithonian "Solnhofener Plattenkalke" in the Museum of Natural History, Ponomarenko" (http://www.landesmuseum.at/pdf_frei_remote/ANNA_87A_0135-0144.pdf). *Ann. Naturhist. Mus. Wien* **87** (1): 135–144. . Retrieved 2011-03-17.

[36] Yan, E. V. (2009). "A New Genus of Elateriform Beetles (Coleoptera, Polyphaga) from the Middle-Late Jurassic of Karatau" (http://fossilinsects.net/pdfs/Yan_2009_PalJ_ElateriformJurassicKaratau.pdf) (PDF). *Paleontological Journal* **43** (1): 78–82. doi:10.1134/S0031030109010080. . Retrieved 2011-03-17.

[37] Tan, J.-J.; , D. Ren, M. Liu (2005). "New Ommatids from the Late Jurassic of western Liaoning, China (Coleoptera: Archostemata)" (http://fossilinsects.net/pdfs/tan_etal_2005.pdf) (PDF). *Insect Science* **12** (3): 207–216. doi:10.1111/j.1005-295X.2005.00026.x. . Retrieved 2011-03-17.

[38] Ponomarenko, A. G. (1997). "New Beetles of the Family Cupedidae from the Mesozoic of Mongolia. Ommatini, Mesocupedini, Priacmini" (http://palaeoentomolog.ru/Publ/PALJ389.pdf) (PDF). *Paleontological Journal* **31** (4): 389–399. . Retrieved 2011-03-17.

[39] Labandeira, C. C.; Sepkoski, J. J. (1993). "Insect diversity in the fossil record". *Science* (Science) **261** (5119): 310–315. doi:10.1126/science.11536548.

[40] GRATSHEV, Vadim G.; ZHERIKHIN, Vladimir V., J. J. (15 Oct., 2003). "Insect diversity in the fossil record" (http://www.isez.pan.krakow.pl/journals/azc_i/pdf/46(suppl)/15.pdf). *Acta zoologica cracoviensia* (Fossil Insects) (261): 129–138. . Retrieved March 15, 2011.

[41] Chang, H.; Zhang, F.; Ren, D. (2008). "A new genus and two new species of fossil Elaterids from the Yixian Formation of Western Liaoning, China (Coleoptera: Elateridae)" (http://202.204.209.200/upload/20080609030648.pdf). *Zootaxa* (1785): 54–62. . Retrieved March 15, 2011.

[42] Orekhovo-Zuyevo, A. V. A. (1993). "Jurassic and Lower Cretaceous Buprestidae (Coleoptera) from Eurasia" (http://www.zin.ru/Animalia/Coleoptera/pdf/Alexeev - jurassic and lc buprestidae 93.pdf). *Paleontological Journal* (27(1A)): 9–34. . Retrieved March 15, 2011.

[43] "New Jewel Beetles (Coleoptera: Buprestidae) from the Cretaceous of Russia, Kazakhstan, and Mongolia" (http://fossilinsects.net/pdfs/Alexeev_2009_PalJ_BuprestidaeCretaceousRussiaKazakhstanMongolia.pdf). *Paleontological Journal* (43): 277–281. 2009. . Retrieved March 15, 2011.

[44] Chin, K.; Gill, B. D. (1996). "Dinosaurs, dung beetles, and conifers; participants in a Cretaceous food web". *PALAIOS* (11): 280–285.

[45] Did dinosaurs have any relation with dung-beetles? (The origin of coprophagy)A. Arillo, V. M. Ortuño, Journal of Natural History 2008, 42(19&20), 1405-1408.

[46] Beutel, R.; F. Haas. (2000). "Phylogenetic relationships of the suborders of Coleoptera (Insecta)" (http://onlinelibrary.wiley.com/doi/10.1111/j.1096-0031.2000.tb00350.x/abstract). *Cladistics* **16**: 103–141. doi:10.1111/j.1096-0031.2000.tb00350.x. . Retrieved 2011-03-18.

[47] J., Kukalová-Peck; J.F. Lawrence. (1993). "Evolution of the hind wing in Coleoptera". *The Canadian Entomologist* **125** (2): 181–258. doi:10.4039/Ent125181-2.

[48] Maddison, David R. (2000 version 11 September 2000 (under construction)). "Coleoptera. Beetle" (http://tolweb.org/coleoptera). *Tree of Life Web Project*. [tolweb.org TolWeb]. . Retrieved 2011-03-18.

[49] Hutchinson, G. Evelyn (1959). "Homage to Santa Rosalia or Why Are There So Many Kinds of Animals?". *The American Naturalist* **93** (870): 145–159. doi:10.1086/282070. JSTOR 2458768.

[50] Mississippi State University. Economic impacts of the boll weevil "History of the Boll Weevil in the United States" (http://www.bollweevil.ext.msstate.edu/webpage_history.htm). Economic impacts of the boll weevil.

[51] "Elm Leaf Beetle" (http://cisr.ucr.edu/elm_leaf_beetle.html). University of California. 2011-May-11. . Retrieved July 17, 2011.

[52] Alyokhin, A., M. Baker, D. Mota-Sanchez, G. Dively, and E. Grafius. 2008. Colorado potato beetle resistance to insecticides. American Journal of Potato Research 85: 395–413.

[53] Adcock, Edward (2005). "Pests - Death watch beelte" (http://www.bodley.ox.ac.uk/dept/preservation/training/pests/watch.htm). *Conservation and collective care*. University of Oxford. . Retrieved July 17, 2011.

[54] Amy R. Remo (September 27, 2007). "Beetles infest coconuts in Manila, 26 provinces" (http://newsinfo.inquirer.net/breakingnews/nation/view_article.php?article_id=91109). Philippine Daily Inquirer. .

[55] "The Mountain Pine Beetle in British Columbia" (http://mpb.cfs.nrcan.gc.ca/biology/introduction_e.html). Natural Resources Canada. August 19, 2008. . Retrieved June 24, 2010.

[56] Anon (5 October 2004). "'Deadly ladybird' sighted in UK" (http://news.bbc.co.uk/1/hi/england/essex/3715120.stm). *BBC News* (BBC). . Retrieved 17 June 2010.
[57] B. Kromp (1999). "Carabid beetles in sustainable agriculture: a review on pest control efficacy, cultivation aspects and enhancement". *Agriculture, Ecosystems and Environment* **74** (1–3): 187–228. doi:10.1016/S0167-8809(99)00037-7.
[58] Brown, J., Scholtz, C.H., Janeau, J-L., Grellier, S. and Podwojewski, P. 2010. Dung beetles (Coleoptera: Scarabaeidae) can improve soil hydrological properties. Applied Soil Ecology 46: 9-16
[59] Losey, John E. and Mace Vaughan (2006). "The Economic Value of Ecological Services Provided by Insects". (http://www.xerces.org/wp-content/uploads/2008/09/economic_value_insects.pdf) *BioScience* 56(4):311-323.
[60] P., Christy. "The Benefit of Dermestid Beetles" (http://www.dailypuppy.com/articles/the-benefit-of-dermestid-beetles/2dc663b4-4292-a1ca-9ab0-648c8378cf7d). The Daily Puppy. . Retrieved July 17, 2011.
[61] Damian Carrington. " Insects could be the key to meeting food needs of growing global population (http://www.guardian.co.uk/environment/2010/aug/01/insects-food-emissions)", *The Guardian* 1 August 2010. Retrieved 27 February 2011.
[62] Ramos-Elorduy, Julieta; Menzel, Peter (1998). *Creepy crawly cuisine: the gourmet guide to edible insects* (http://books.google.com/?id=Q7f1LkFz11gC). Inner Traditions / Bear & Company. pp. 150. ISBN 9780892817474. . Retrieved 2010-04-02.
[63] Life cycle of the rounded jewel beetles, Sternocera spp. วงจรชีวิตของแมลงทับกลมใช้เวลานานถึง 2 ปี - Siam Insect Zoo-Museum (http://www.malaeng.com/blog/?p=4262)
[64] Michael A. Ivie (2002). "105. Zopheridae". In Ross H. Arnett & Michael Charles Thomas. *American Beetles: Polyphaga: Scarabaeoidea through Curculionoidea*. Volume 2 of American Beetles. CRC Press. pp. 457–462. ISBN 9780849309540.
[65] Zabludoff, Marc (2008). *Beetles*. Malaysia: Michelle Bison. pp. 14-17. ISBN 9780761425328.
[66] Dollinger, André (January 2002). "Ancient Egyptian bestiary: Insects" (http://www.reshafim.org.il/ad/egypt/bestiary/insects.htm). . Retrieved July 19, 2011.
[67] "Isis and Osiris" (http://penelope.uchicago.edu/Thayer/E/Roman/Texts/Plutarch/Moralia/Isis_and_Osiris*/E.html#ref373), *Moralia*, in volume V of the Loeb Classical Library edition, 1936, now in the public domain. Retrieved on 2007-08-02.
[68] Morales-Correa, Ben (2006). "Egyptian Symbols" (http://www.all-about-egypt.com/egyptian-symbols.html). All-About-Egypt. . Retrieved July 19, 2011.

External links

Global

- Coleoptera (http://tolweb.org/Coleoptera/8221) from the Tree of Life Web Project
- List of major Beetle collections (http://www.fond4beetles.com/collections.html)
- Beetles and coleopterologists (http://www.zin.ru/Animalia/Coleoptera/eng/index.htm)
- (German) *Käfer der Welt* (http://www.kaefer-der-welt.de/index.htm)
- Coleop-Terra (http://www.coleop-terra.com/about.html)
- Beetles – Coleoptera (http://www.insectlifeforms.com/Orders736/BEETLES_COLEOPTERA_60502060_736.aspx)
- Beetle larvae (http://www.flickr.com/photos/25258027@N02/sets/72157604361795999/)
- Beetle images (http://www.flickr.com/photos/coleoptera-us/sets/72157607363771409/)

Regional

- Gallery of European beetles (http://www.koleopterologie.de/gallery)
- Poland beetles (http://www.colpolon.biol.uni.wroc.pl/lista rodzin.htm)
- Identification keys to some British beetles (http://sites.google.com/site/mikesinsectkeys)
- North American Beetles (http://bugguide.net/node/view/60)
- Beetles of North America (http://www.cirrusimage.com/beetles.htm)
- Texas beetle information (http://www.texasento.net/beetles.htm)
- The Beetle Ring (http://www.naturalworlds.org/beetlering/index.htm)
- Beetles of Africa (http://www.beetlesofafrica.com/index-old.asp)
- Beetles of Mauritius (http://mauritiusbeetles.myspecies.info)
- Southeast Asian beetles (http://www.beetle-diversity.com)

bjn:Karariang pfl:Käwwer

Oryzaephilus surinamensis

Oryzaephilus surinamensis	
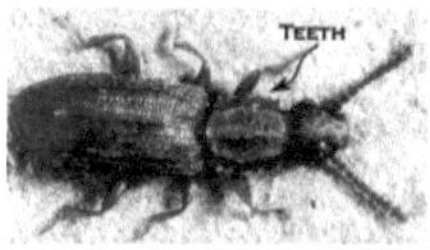	
Scientific classification	
Kingdom:	Animalia
Phylum:	Arthropoda
Class:	Insecta
Order:	Coleoptera
Family:	Silvanidae
Genus:	*Oryzaephilus*
Species:	***O. surinamensis***
Binomial name	
Oryzaephilus surinamensis (Linnaeus, 1758)	

The **sawtoothed grain beetle**, ***Oryzaephilus surinamensis***, is a slender, dark brown beetle 2.4–3 mm in size , with characteristic "teeth" running down the side of the prothorax. The sawtoothed grain beetle is a primary pest, in the UK, that attacks damaged grain. Carolus Linnaeus, the Swedish naturalist who defined taxonomic arrangements for the animal, plant, and mineral kingdoms acquired samples of the sawtoothed grain beetle from Suriname (Dutch Guiana) and named the species "*Oryzaephilus surinamensis*" because of this. (It is Latin for "rice-lover from Surinam".)

A relative of the sawtoothed grain beetle of the same genus, the merchant grain beetle, is more likely to be found in domestic dwellings since it requires higher more stable temperatures and can be found infesting biscuits and fruit and nut products. The adults can fly but rarely do. They can be differentiated by looking at the distance of the eye to the prothorax: with the merchant grain beetle it is less than half the vertical diameter of the eye.

The sawtoothed grain beetle lays its eggs loosely on foodstuffs at the rate of 6–10 per day, with total being 370 per female. The larvae are to be found within the mass of the foodstuff in the top centimetre or two. As mentioned above damaged cereal is entered through broken kernels, and the larvae feed on the germ, causing damage by reducing the percentage of grains which will germinate.

The total life cycle is 20–80 days at 18–37 °C (64–99 °F).

The sawtoothed grain beetle can survive in unheated warehouses in the UK, and is considered the most important pest of home grown grain in the UK. The beetles survive the winter hiding in cracks and infest new stocks of grain the following year. Both insects are distributed throughout the world and regularly imported into the UK in unprocessed cereals, oil cakes, and from grains from other stores.

Treatment in bulk storage situations normally is carried out by fumigation, and in domestic situations finding the infested material, disposing of it and carrying out a residual spray to kill off stragglers.

Flea

Flea Temporal range: [1]	
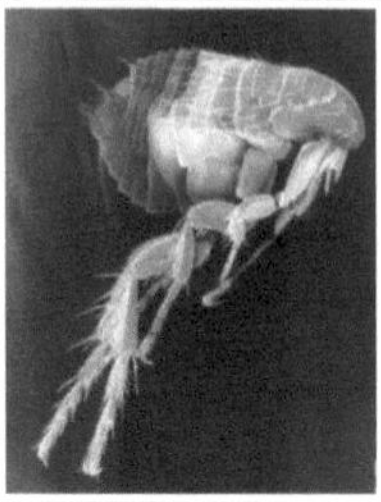	
Scanning electron microscope (SEM) depiction of a flea	
Scientific classification	
Kingdom:	Animalia
Phylum:	Arthropoda
Class:	Insecta
Subclass:	Pterygota
Infraclass:	Neoptera
Superorder:	Endopterygota
Order:	**Siphonaptera** Latreille, 1825
Infraorders	
Ceratophyllomorpha Hystrichopsyllomorpha Pulicomorpha Pygiopsyllomorpha	
Synonyms	
Aphaniptera	

Flea is the common name for insects of the order **Siphonaptera** which are wingless insects with mouthparts adapted for piercing skin and sucking blood. Fleas are external parasites, living by hematophagy off the blood of mammals (including bats and humans) and birds.

Some flea species include:

- Cat flea (*Ctenocephalides felis*)
- Dog flea (*Ctenocephalides canis*)
- Human flea (*Pulex irritans*)
- Moorhen flea (*Dasypsyllus gallinulae*)
- Northern rat flea (*Nosopsyllus fasciatus*)
- Oriental rat flea (*Xenopsylla cheopis*)

Over 2,000 species have been described worldwide.[2]

Morphology and behavior

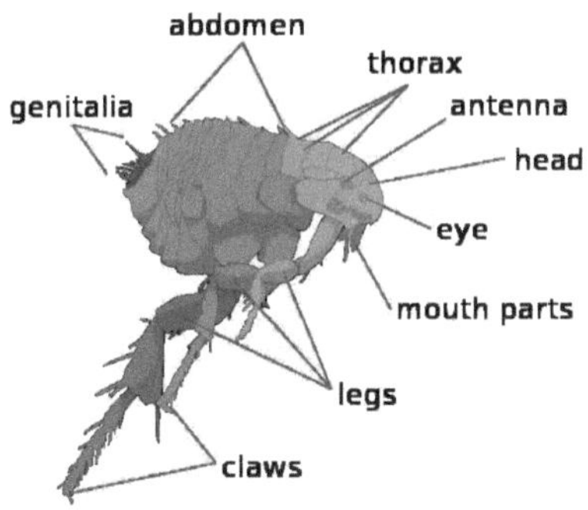

Diagram of a Flea

Fleas are insects (1/16 to 1/8-inch (1.5 to 3.3 mm) long), agile, usually dark colored (for example, the reddish-brown of the cat flea), wingless insects with tube-like mouth-parts adapted to feeding on the blood of their hosts. Their legs are long, the hind pair well adapted for jumping: a flea can jump vertically up to 7 inches (18 cm) and horizontally up to 13 inches (33 cm).[3] This is around 200 times their own body length, making the flea one of the best jumpers of all known animals (relative to body size), second only to the froghopper. According to an article in *Science News*, "researchers with the University of Cambridge in England have shown that fleas take off from their tibiae and tarsi—the insect equivalent of feet—and not their trochantera, or knees. The researchers report their conclusion in the March 1 Journal of Experimental Biology."[4] It has been known that fleas do not use muscle power but energy stored in a protein named resilin but the researchers used high-speed video technology and mathematical models to discover where the spring action actually happens. Their bodies are laterally compressed (human anatomical terms), permitting easy movement through the hairs or feathers on the host's body (or in the case of humans, under clothing). The flea body is hard, polished, and covered with many hairs and short spines directed backward,[5] which also assist its movements on the host. The tough body is able to withstand great pressure, likely an adaptation to survive attempts to eliminate them by mashing or scratching. Even hard squeezing between the fingers is normally insufficient to kill a flea. It is possible to eliminate them by pressing individual fleas with adhesive tape or softened beeswax (or "cheese" wax) or by rolling a flea briskly between the fingers to disable it then crushing it between the fingernails. Fleas also can be drowned in water and may not survive direct contact with anti-flea pesticides.

Fleas lay tiny white oval-shaped eggs better viewed through a loupe. The larva is small, pale, has bristles covering its worm-like body, lacks eyes, and has mouthparts adapted to chewing. The larvae feed on various organic matter, especially the feces of mature fleas. The adult flea's diet consists solely of fresh blood.[6] In the pupal phase, the larva is enclosed in a silken, debris-covered cocoon.

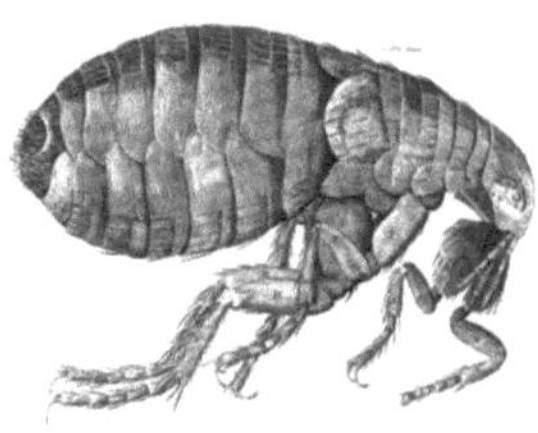

Hooke's drawing of a flea in *Micrographia*

Life cycle and habitat

Fleas are holometabolous insects, going through the four life cycle stages of egg, larva, pupa, and imago (adult). The flea life cycle begins when the female lays after feeding. Adult fleas must feed on blood before they can become capable of reproduction.[5] Eggs are laid in batches of up to 20 or so, usually on the host itself, which means that the eggs can easily roll onto the ground. Because of this, areas where the host rests and sleeps become one of the primary habitats of eggs and developing fleas. The eggs take around two days to two weeks to hatch.[3]

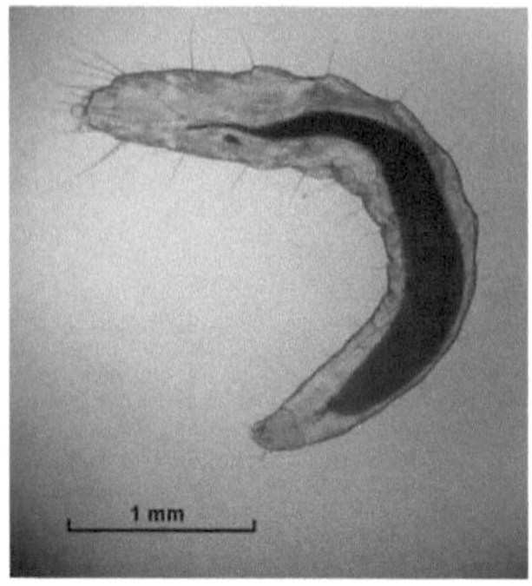

Micrograph of a flea larva.

Fleas pass through a complete life cycle consisting of egg, larva, pupa and adult. A typical flea population consists of 50% eggs, 35% larvae, 10% pupae and 5% adults. Completion of the life cycle from egg to adult varies from two weeks to eight months depending on the temperature, humidity, food, and species. Normally after a blood meal, the female flea lays about 45 to 50 eggs per day up to 600 in a lifetime usually on the host (dogs, cats, rats, rabbits, mice, squirrels, chipmunks, raccoons, opossums, foxes, chickens, humans, etc.). Eggs loosely laid in the hair coat, drop out most anywhere especially where the host rests, sleeps or nests (rugs, carpets, upholstered furniture, cat or dog boxes, kennels, sand boxes, etc.). Eggs hatch in two days to two weeks into larvae found indoors in floor cracks & crevices, along baseboards, under rug edges and in furniture or beds. Outdoor development occurs in sandy gravel soils (moist sand boxes, dirt crawlspace under the house, under shrubs, etc.) where the pet may rest or sleep. Sand and gravel are very suitable for larval development which is the reason fleas are erroneously called "sand fleas."

Larvae are blind, avoid light, pass through three larval instars and take a week to several months to develop. Their food consists of digested blood from adult flea feces, dead skin, hair, feathers, and other organic debris. (Larvae do not suck blood.) Pupa mature to adulthood within a silken cocoon woven by the larva to which pet hair, carpet fiber, dust, grass cuttings, and other debris adheres. In about five to fourteen days, adult fleas emerge or may remain resting in the cocoon until the detection of vibration (pet and people movement), pressure (host animal lying down on them), heat, noise, or carbon dioxide (meaning a potential blood source is near). Most fleas overwinter in the larval or pupal stage with survival and growth best during warm, moist winters and spring.

Adult fleas cannot lay eggs without a blood meal, but may live for one year without feeding. There is often a desperate need for flea control after a family has returned from a long vacation. The house has been empty with no cat or dog around for fleas to feed on. When the family and pets are gone, flea eggs hatch and larvae pupate. The adult fleas fully developed inside the pupal cocoon remain in a kind of "limbo" for a long time until a blood source is near. The family returning from vacation is immediately attacked by waiting hungry hordes of fleas. (In just 30 days, 10 female fleas under ideal conditions can multiply to over a quarter million fleas of different life stages.)

Newly emerged adult fleas live only about one week if a blood meal is not obtained. However, completely developed adult fleas can live for several months without eating, so long as they do not emerge from their puparia. Optimum temperatures for the flea's life cycle are 70°F to 85°F (21°C to 30°C) and optimum humidity is 70%.[7]

Classification

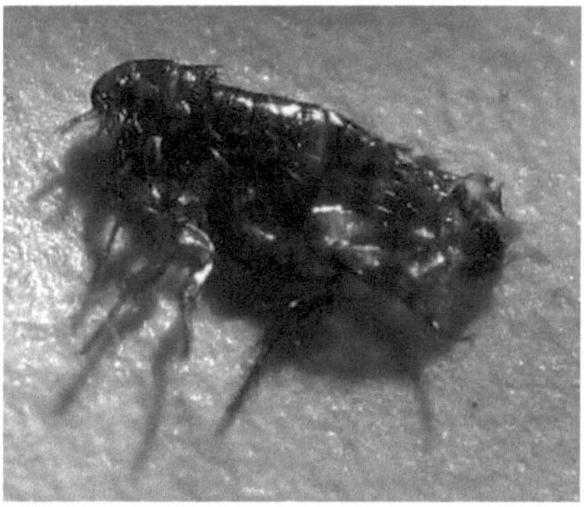

Cat flea

Fleas are related to mecoptera,[8] winged insects with good eyesight. The flightless boreid (snow scorpionfly) with its rudimentary wings seems to be close to the common ancestor of the 2000 or so known varieties of flea, which split off in many directions around 160 million years ago.[8] Their evolution continued to produce adaptations for their specialized parasitic niche, such that they now have no wings and their eyes are covered over. The large number of flea species may be attributed to the wide variety of host species they feed on, which provides so many specific ecological niches to adapt to.

In the past, it was most commonly supposed that fleas had evolved from the flies (Diptera), based on similarities of the larvae. (Some authorities use the name **Aphaniptera** because it is older, but names above family rank need not follow the ICZN rules of priority, so most taxonomists use the more familiar name). Genetic and morphological evidence indicates that they are descendants of the Scorpionfly family Boreidae, which are also flightless; accordingly it is possible that they will eventually be reclassified as a suborder within the Mecoptera. In any case, all these groups seem to represent a clade of closely related insect lineages, for which the names Mecopteroidea and Antliophora have been proposed.

Flea systematics are not entirely fixed. While, compared to many other insect groups, fleas have been studied and classified fairly thoroughly, details still remain to be learned about the evolutionary relationships among the different flea lineages.

- Infraorder Pulicomorpha
  - Superfamily Pulicoidea
    - Family Hectopsyllidae—sticktight and chigoe fleas ("chiggers" of Latin America)
    - Family Pulicidae—common fleas
  - Superfamily Malacopsylloidea
    - Family Malacopsyllidae
    - Family Rhopalopsyllidae—hosts
    - Family Vermipsyllidae—hosts: carnivores
  - Superfamily Coptopsylloidea
    - Family Coptopsyllidae
  - Superfamily Ancistropsylloidea
    - Family Ancistropsyllidae
- Infraorder Pygiopsyllomorpha
  - Superfamily Pygiopsylloidea
    - Family Lycopsyllidae
    - Family Pygiopsyllidae
    - Family Stivaliidae
- Infraorder Hystrichopsyllomorpha
  - Superfamily Hystrichopsylloidea
    - Family Hystrichopsyllidae—hosts: rats and mice. Includes Ctenopsyllidae, Amphipsyllidae

- Family Chimaeropsyllidae
- Superfamily Macropsylloidea
  - Family Macropsyllidae
- Superfamily Stephanocircidoidea
  - Family Stephanocircidae

Relationship with host

Fleas attack a wide variety of warm-blooded vertebrates including dogs, cats, humans, chickens, rabbits, squirrels, rats, ferrets, and mice.

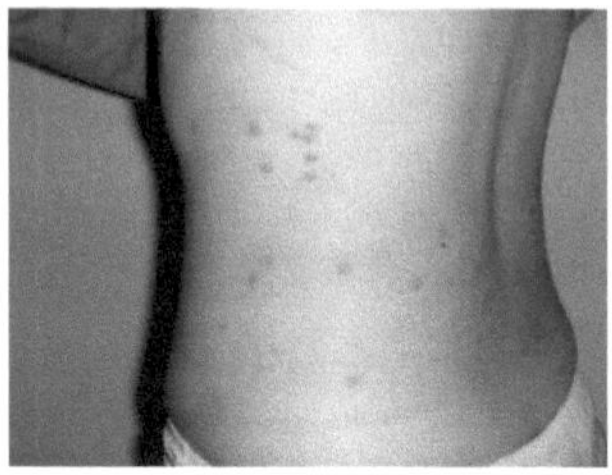

Flea bites on the back of a human

Direct effects of bites

Fleas are not only a nuisance to humans and their pets, but can cause medical problems including flea allergy dermatitis (FAD), secondary skin irritations and, in extreme cases, anemia, tapeworms, stomach flu. Fleas can transmit murine typhus (endemic typhus) fever among animals and from animal to humans. Also, fleas can transmit bubonic plague and any other disease from human to rodent and from rodent to humans. Tapeworms can infest in humans in severe cases. Although bites are rarely felt, it is the resulting irritation caused by the flea salivary secretions that varies among individuals. Some may witness a severe reaction (general rash or inflammation) resulting in secondary infections caused by scratching the irritated skin area. Others may show no reaction or irritation acquired after repeated bites over several weeks or months. Most bites usually found on the ankles and legs may cause irritation or pain lasting a few minutes, hours or days depending on one's sensitivity. The typical reaction to the bite is the formation of a small, hard, red, slightly raised (swollen) itching spot. There is a single puncture point in the center of each spot. (Ants and spiders leave two marks when they bite. Mosquitoes, bees, wasps and bedbugs cause a large swelling or welt.)[9] :126

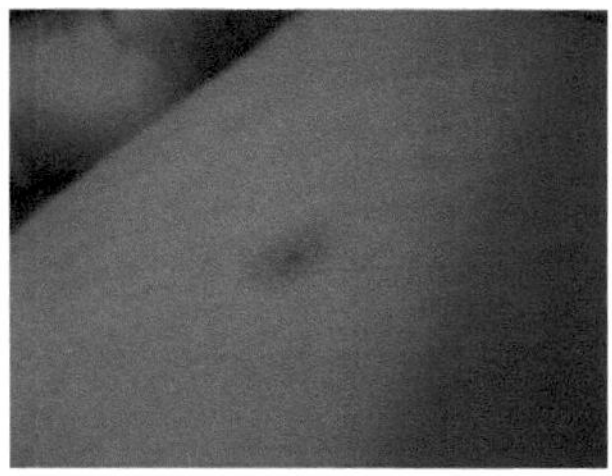

Flea bite on the waist of a human with no reaction

As a vector

Besides the problems posed by the creature itself, fleas can also act as a vector for disease. Fleas transmit not only a variety of viral, bacterial and rickettsial diseases to humans and other animals, but also protozoans and helminths.[10] :72–73

- bacteria: Murine or endemic typhus.[9] :124 Fleas have helped cause epidemics by transmitting diseases such as the bubonic plague between rodents and humans by carrying *Yersinia pestis* bacteria.[11] Fleas can transmit *Yersinia pestis*, *Rickettsia typhi*, *Rickettsia felis*, and *Bartonella henselae*.
- virus: myxomatosis.[10] :73
- helminth: infestation of *Hymenolepiasis* tapeworm.[12]
- protozoa: Trypanosome protozoans such as those of the subgenus *Herpetosoma*, use a variety of flea species opportunistically as vectors.[10] :74

Fleas that specialize as parasites on specific mammals may use other mammals as hosts; therefore humans are susceptible to the predation of more than one species of flea.[13]

A myth and misconception of Fleas carring the HIV/AIDS have been debunked by the Center for Disease and Prevention (CDC 2003) in which they cannot carry the virus and spread it to other humans.

Flea treatments

For humans

Fleas can become settled in a person's hair in less than ten minutes and cause soreness and itching. The itching associated with flea bites can be treated with anti-itch creams, usually antihistamines or hydrocortisone.[14] Calamine lotion has been shown to be effective for itching.[15]

For pets

Modern flea control is approached using Integrated Pest Management (IPM) protocols at the host (pet) level. IPM is achieved by targeting fleas during at least two separate life stages, with at least two separate molecules. This is typically achieved using an adulticide to kill adult fleas and an insect development inhibitor (IDI), like lufenuron, or insect growth regulator (IGR), like methoprene, to prevent development of immature stages. Flea adults, larvae, or eggs can be controlled with insecticides. Lufenuron is a veterinary preparation (Program) that attacks the larval flea's ability to produce chitin but does not kill fleas. Flea medicines need to be used with care because many of them also affect mammals.

Flea and tick repellant powder being applied to a dog

Cedar oil, a non-toxic natural substance, has been proven effective in the eradication of infestations in pets. Cedar oil is being used to treat sand-flea infestation of US Military forces in the Gulf.[16]

Since more than three quarters of a flea's life is spent somewhere other than on the host animal, it is not adequate to treat only the host; it is important also to treat the host's environment. Thorough vacuuming, washing linens in hot water, and treating all hosts in the immediate environment (the entire household, for example) are essential and if possible on a regular basis.[17]

Contemporary commercial products for the topical treatment of flea infestations on pets contain pesticides such as imidacloprid, permethrin, and (S)-methoprene. All flea control products are recommended to be used at least half a year because the lifecycle of flea and tick can last to up to 6 months, and by using one of the flea and tick control products for so long, the infestation is highly prevented and, in the end, stopped. Although all these products are effective in fighting against flea and tick infestations, they have different active ingredients and, because cats cannot metabolize some of the compounds of the product, care must be taken in their use.

For the home

Combating a flea infestation in the home takes patience because for every flea found on an animal, there could be many more developing in the home. A spot-on insecticide will kill the fleas on the pet and in turn the pet itself will be a roving flea trap and mop up newly hatched fleas. The environment should be treated with a fogger or spray insecticide containing an insect growth regulator, such as pyriproxyfen or methoprene to kill eggs and pupae, which are quite resistant against insecticides. Frequent vacuuming is also helpful, but the vacuum bag must be disposed of immediately afterwards.

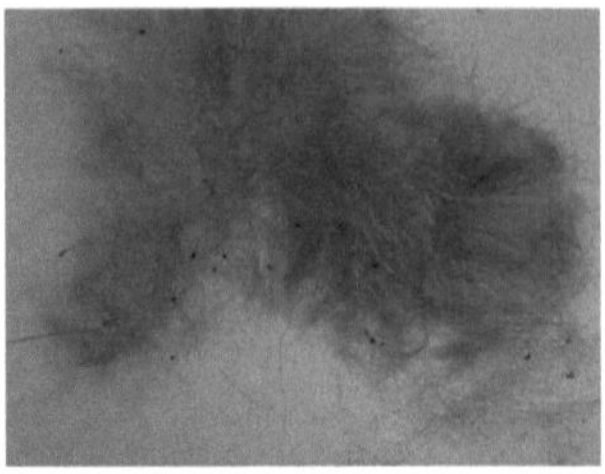

Flea "dirt" in the fur of a cat is actually excess blood from the host consumed by the adult flea and passed as feces

Diatomaceous earth can also be used as a home flea treatment in lieu of acetylcholinesterase inhibitory treatments or insecticides which carry with them a risk of poisoning for both humans and animals. However, diatomaceous earth is at least potentially dangerous to pets and people when inhaled, so care in use is recommended.

Dried pennyroyal has been suggested as a natural flea control,[18] but is not recommended in homes with pets due to its high toxicity to mammals.[19]

Borax is sold as a "Natural Laundry Booster" and can also be used as another home treatment for flea infestations. Borax contains boric acid which kills fleas by dehydrating them, but its safety for pets is untested.[20]

Using dehumidifiers with air conditioning and vacuuming all may interrupt the flea life cycle. Humidity is critical to flea survival. Eggs need relative humidity of at least 70–75% to hatch, and larvae need at least 50% humidity to survive. In humid areas, about 20% of the eggs survive to adulthood; in arid areas, less than 5% complete the cycle.[21] Fleas thrive at higher temperatures, but need 70° to 90°F (21° to 32°C) to survive. Lower temperatures slow down or completely interrupt the flea life-cycle. A laboratory study done at the University of California showed that vacuuming catches about 96% of adult fleas. A combination of controlled humidity, temperature, and vacuuming should eliminate fleas from an environment, and altering even one of these environmental factors may be enough to drastically lower and eliminate an infestation.

See also

- Pulicosis (Flea bites)
- Anti-itch drug
- Bubonic plague
- Cat flea
- *Echidnophaga gallinacea*
- Use of DNA in forensic entomology

References

[1] Hoell, H.V., Doyen, J.T. & Purcell, A.H. (1998). *Introduction to Insect Biology and Diversity, 2nd ed.*. Oxford University Press. p. 320. ISBN 0-19-510033-6.

[2] Fleas: What They Are, What To Do (http://edis.ifas.ufl.edu/ig132) D. L. Richman and P. G. Koehler, University of Florida IFAS Extension. Accessed 10 December 2010

[3] Fleas, HYG-2081-97 (http://ohioline.osu.edu/hyg-fact/2000/2081.html) William F. Lyon, Ohio State University entomology page. Accessed 28 December 2006

[4] http://www.sciencenews.org/view/generic/id/69664/title/Fleas_leap_from_feet%2C_not_knees

[5] Fleas (http://edis.ifas.ufl.edu/IG087). P.G. Koehler and F. M. Oi. Printed July 1993, revised February 2003. Provided by the University of Florida

[6] Order Siphonaptera – Fleas – BugGuide.Net (http://bugguide.net/node/view/7040) Accessed 28 December 2006

[7] Piper, Ross (2007), *Extraordinary Animals: An Encyclopedia of Curious and Unusual Animals*, Greenwood Press.

[8] Grimaldi, D. and Engel, M.S. (2005). *Evolution of the Insects*. Cambridge University Press. ISBN 0-521-82149-5.

[9] Mullen, Gary R.; Mullen, Gary; Durden, Lance (2009). *Medical and Veterinary Entomology* (http://books.google.com/?id=6R1v9o-ual4C). Academic Press. pp. 637. ISBN 978-0-12-372500-4. . Retrieved 2010-04-09.

[10] Krasnov, Boris R. (2008). *Functional and evolutionary ecology of fleas: a model for ecological parasitology* (http://books.google.com/?id=tkvr8h8zxFwC). Cambridge University Press. pp. 593. ISBN 978-0-521-88277-4. . Retrieved 2010-04-09.

[11] [|Sherman, David M. (http://books.google.co.in/books?q=+inauthor:"David+M.+Sherman"&lr=&as_drrb_is=q&as_minm_is=0&as_miny_is=&as_maxm_is=0&as_maxy_is=&num=20&as_brr=3&source=gbs_metadata_r&cad=10)] (2002). *Tending animals in the global village: a guide to international veterinary medicine* (http://books.google.com/?id=ax_lLPG634AC). Wiley-Blackwell. p. 209.

ISBN 978-0-683-18051-0. . Retrieved 2010-04-09.

[12] Stein, Ernst (2003). *Anorectal and colon diseases: textbook and color atlas of proctology* (http://books.google.com/?id=yrlD7i4A68IC). Springer. p. 478. ISBN 978-3-540-43039-1. . Retrieved 2010-04-09.

[13] Barnes, Ethne (2007). *Diseases and Human Evolution* (http://books.google.com/?id=wnlB7EMtqFYC). UNM Press. p. 253. ISBN 978-0-8263-3066-6. . Retrieved 2010-04-09.

[14] Scott, Susan; Thomas, Craig (2000). *Pests of Paradise: First Aid and Medical Treatment of Injuries from Hawaii's Animals* (http://books.google.com/?id=0c3xvwaJNJkC). University of Hawaii Press. p. 58. ISBN 978-0-8248-2252-1. . Retrieved 2010-04-08.

[15] Jacoby, David B.; Youngson, R.M. (2004). *Encyclopedia of Family Health* (http://books.google.com/?id=NC-87V00OJwC). Marshall Cavendish. p. 647. ISBN 978-0-7614-7486-9. . Retrieved 2010-04-08.

[16] "Lights Out For Bed Bugs as "BEST YET" Cedar Oil Solution Solves International Bed Bug Dilemma" (http://www.prweb.com/releases/2008/11/prweb1659224.htm). http://www.prweb.com/. 2008-11-25. . Retrieved 2010-04-08.

[17] ntp.niehs.nih.gov/ntp/htdocs/Chem.../cedarwood_oil.pdf

[18] Natural Flea Control (http://www.motherearthnews.com/Livestock-and-Farming/1987-07-01/Natural-Flea-Control.aspx), 1987, Mother Earth News (http://www.motherearthnews.com), Retrieved 2007-12-18

[19] *Warnings About Essential Oils* (http://bitsandbrew.com/warning1.htm)

[20] Fleas (http://www.pesticide.org/fleas.html)

[21] Fleas Management Guidelines (http://www.ipm.ucdavis.edu/PMG/PESTNOTES/pn7419.html). Univ. of California Division of Agriculture and Natural Resources, Nov. 2000.

External links

- Flea (http://www.dmoz.org/Science/Biology/Flora_and_Fauna/Animalia/Arthropoda/Insecta/Fleas//) at the Open Directory Project
- Flea information (http://pestworld.org/for-consumers/Pest-Guide/Pest/Flea) highlighting habits, habitat, threats and prevention tips
- Integrated Flea Control from University of Nebraska-Lincoln Extension in Lancaster County (http://lancaster.unl.edu/pest/resources/IntegratedFlea.shtml)
- Flea Pest Control Information (http://npic.orst.edu/pest2.htm#fleas)—National Pesticide Information Center
- Advice about fleas (http://parents.berkeley.edu/advice/health/fleas.html) from the Berkeley Parents Network

Drosophila melanogaster

Drosophila melanogaster	
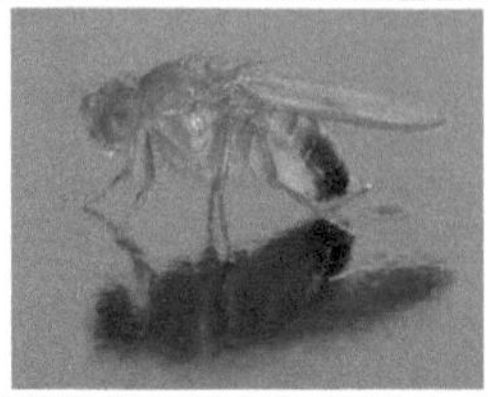	
Male *Drosophila melanogaster*	
Scientific classification	
Kingdom:	Animalia
Phylum:	Arthropoda
Class:	Insecta
Order:	Diptera
Family:	Drosophilidae
Genus:	*Drosophila*
Subgenus:	*Sophophora*
Species group:	*melanogaster* group
Species subgroup:	*melanogaster* subgroup
Species complex:	*melanogaster* complex
Species:	***D. melanogaster***
Binomial name	
Drosophila melanogaster Meigen, 1830[1]	

Drosophila melanogaster (Greek for *dark-bellied dew lover* : δρόσος = dew, φίλος = intimate friend, lover, μέλας = dark-coloured, γαστήρ = belly [2]) is a species of Diptera, or the order of flies, in the family Drosophilidae. The species is commonly known as the **common fruit fly** or **vinegar fly**. Starting from Charles W. Woodworth, this species is one of the most commonly used model organisms in biology, including studies in genetics, physiology, microbial pathogenesis and life history evolution because they are easy to take care of, breed quickly, and lay many eggs.[3]

Flies belonging to the family Tephritidae are also called fruit flies, which can lead to confusion, especially in Australia where the term fruit fly refers to the Tephritidae, economic pests in fruit production.

Physical appearance

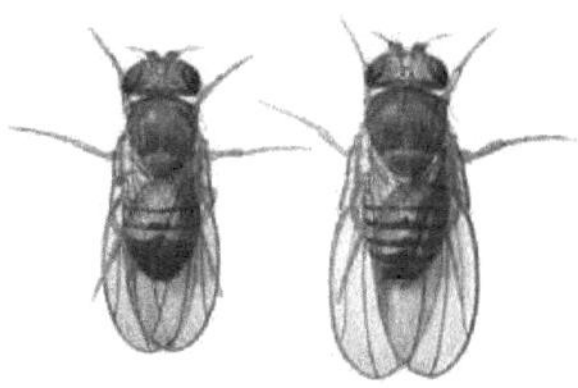

Male (left) and female *D. melanogaster*

Wildtype fruit flies have brick red eyes, are yellow-brown in color, and have transverse black rings across their abdomen. They exhibit sexual dimorphism: females are about 2.5 millimeters (0.098 in) long; males are slightly smaller and the back of their bodies is darker. Males are easily distinguished from females based on color differences, with a distinct black patch at the abdomen, less noticeable in recently emerged flies (see fig), and the sexcombs (a row of dark bristles on the tarsus of the first leg). Furthermore, males have a cluster of spiky hairs (claspers) surrounding the reproducing parts used to attach to the female during mating. There are extensive images at FlyBase.[4]

Life cycle and reproduction

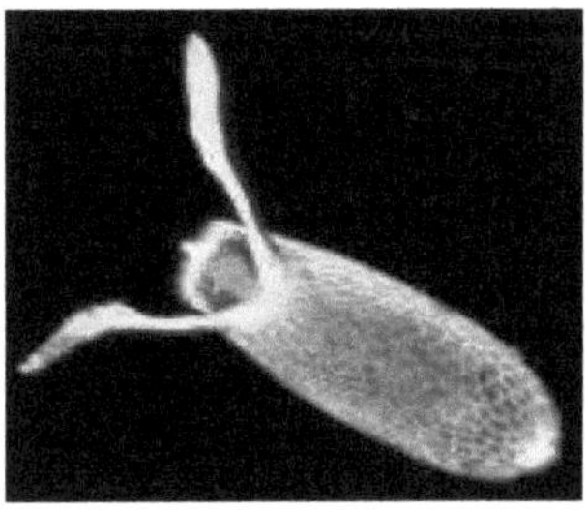

Egg of *D. melanogaster*

The *D. melanogaster* lifespan is about 30 days at 29 °C (84 °F).

The developmental period for *Drosophila melanogaster* varies with temperature, as with many ectothermic species. The shortest development time (egg to adult), 7 days, is achieved at 28 °C (82 °F).[5] [6] Development times increase at higher temperatures (11 days at 30 °C or 86 °F) due to heat stress. Under ideal conditions, the development time at 25 °C (77 °F) is 8.5 days,[5] [6] [7] at 18 °C (64 °F) it takes 19 days[5] [6] and at 12 °C (54 °F) it takes over 50 days.[5] [6] Under crowded conditions, development time increases,[8] while the emerging flies are smaller.[8] [9] Females lay some 400 eggs (embryos), about five at a time, into rotting fruit or other suitable material such as decaying mushrooms and sap fluxes. The eggs, which are about 0.5 millimetres long, hatch after 12–15 hours (at 25 °C or 77 °F).[5] [6] The resulting larvae grow for about 4 days (at 25 °C) while molting twice (into 2nd- and 3rd-instar larvae), at about 24 and 48 h after hatching.[5] [6] During this time, they feed on the microorganisms that decompose the fruit, as well as on the sugar of the fruit itself. Then the larvae encapsulate in the puparium and undergo a four-day-long metamorphosis (at 25 °C), after which the adults eclose (emerge).[5] [6]

Mating fruit flies. Note the sex combs on the forelegs of the male (insert)

Females become receptive to courting males at about 8–12 hours after emergence.[10] Males perform a sequence of five behavioral patterns to court females. First, males orient themselves while playing a courtship song by horizontally extending and vibrating their wings. Soon after, the male positions itself at the rear of the female's abdomen in a low posture to tap and lick the female genitalia. Finally, the male curls its abdomen, and attempts copulation. Females can reject males by moving away, kicking and extruding their ovipositor. Copulation lasts around 15–20 minutes,[11] during which males transfer a few hundred very long (1.76 mm) sperm cells in seminal fluid to the female.[12] Females store the sperm in a tubular receptacle and in two mushroom-shaped spermathecae, sperm from multiple matings compete for fertilization. A last male precedence is believed to exist in which the last male to mate with a female sires approximately 80% of her offspring. This precedence was found to occur through displacement and incapacitation.[13] The displacement is attributed to sperm handling by the female fly as multiple matings are conducted and is most significant during the first 1–2 days after copulation. Displacement from the seminal receptacle is more significant than displacement from the spermathecae.[13] Incapacitation of first male sperm by second male sperm becomes significant 2–7 days after copulation. The seminal fluid of the second male is believed to be responsible for this incapacitation mechanism (without removal of first male sperm) which takes effect before fertilization occurs.[13] The delay in effectiveness of the incapacitation mechanism is believed to be a protective mechanism that prevents a male fly from incapacitating its own sperm should it mate with the same female fly repetitively.[13]

History of use in genetic analysis

Lab cultures

Drosophila melanogaster was among the first organisms used for genetic analysis, and today it is one of the most widely used and genetically best-known of all eukaryotic organisms. All organisms use common genetic systems; therefore, comprehending processes such as transcription and replication in fruit flies helps in understanding these processes in other eukaryotes, including humans.[14]

Charles W. Woodworth is credited with being the first to breed *Drosophila* in quantity and for suggesting to W. E. Castle that they might be used for genetic research during his time at Harvard University.

Thomas Hunt Morgan began using fruit flies in experimental studies of heredity at Columbia University in 1910. His laboratory was located on the top floor of Schermerhorn Hall, which became known as the Fly Room. The Fly Room was cramped with eight desks, each occupied by students and their experiments. They started off experiments using milk bottles to rear the fruit flies and handheld lenses for observing their traits. The lenses were later replaced by microscopes, which enhanced their observations. The Fly Room was the source of some of the most important research in the history of biology. Morgan and his students eventually elucidated many basic principles of heredity, including sex-linked inheritance, epistasis, multiple alleles, and gene mapping.[14]

"Thomas Hunt Morgan and colleagues extended Mendel's work by describing X-linked inheritance and by showing that genes located on the same chromosome do not show independent assortment. Studies of X-linked traits helped confirm that genes are found on chromosomes, while studies of linked traits led to the first maps showing the locations of genetic loci on chromosomes" (Freman 214). The first maps of *Drosophila* chromosomes were completed by Alfred Sturtevant.

Model organism in genetics

Drosophila melanogaster is one of the most studied organisms in biological research, particularly in genetics and developmental biology. There are several reasons:

D. melanogaster types (clockwise): brown eyes with black body, cinnabar eyes, sepia eyes with ebony body, vermilion eyes, white eyes, and wild-type eyes with yellow body

- Its care and culture requires little equipment and uses little space even when using large cultures, and the overall cost is low.
- It is small and easy to grow in the laboratory and their morphology is easy to identify once they are anesthetized (usually with ether, carbon dioxide gas, by cooling them, or with products like FlyNap)
- It has a short generation time (about 10 days at room temperature) so several generations can be studied within a few weeks.
- It has a high fecundity (females lay up to 100 eggs per day, and perhaps 2000 in a lifetime).[3]
- Males and females are readily distinguished and virgin females are easily isolated, facilitating genetic crossing.
- The mature larvae show giant chromosomes in the salivary glands called polytene chromosomes—"puffs" indicate regions of transcription and hence gene activity.
- It has only four pairs of chromosomes: three autosomes, and one sex chromosome.
- Males do not show meiotic recombination, facilitating genetic studies.
- Recessive lethal "balancer chromosomes" carrying visible genetic markers can be used to keep stocks of lethal alleles in a heterozygous state without recombination due to multiple inversions in the balancer.
- Genetic transformation techniques have been available since 1987.
- Its complete genome was sequenced and first published in 2000.[15]

Genetic markers

Genetic markers are commonly used in *Drosophila* research, for example within balancer chromosomes or P-element inserts, and most phenotypes are easily identifiable either with the naked eye or under a microscope. In the list of example common markers below, the allele symbol is followed by the name of the gene affected and a description of its phenotype. *(Note: Recessive alleles are in lower case, while dominant alleles are capitalised.)*

- Cy^1: curly; The wings curve away from the body, flight may be somewhat impaired.
- e^1: ebony; Black body and wings (heterozygotes are also visibly darker than wild type).
- Sb^1: stubble; Hairs are shorter and thicker than wild type.
- w^1: white; Eyes lack pigmentation and appear white, vision may be somewhat impaired.
- y^1: yellow; Body pigmentation and wings appear yellow.

Drosophila genes are traditionally named after the phenotype they cause when mutated. For example, the absence of a particular gene in *Drosophila* will result in a mutant embryo that does not develop a heart. Scientists have thus called this gene *tinman*, named after the Oz character of the same name.[16] This system of nomenclature results in a wider range of gene names than in other organisms.

Genome

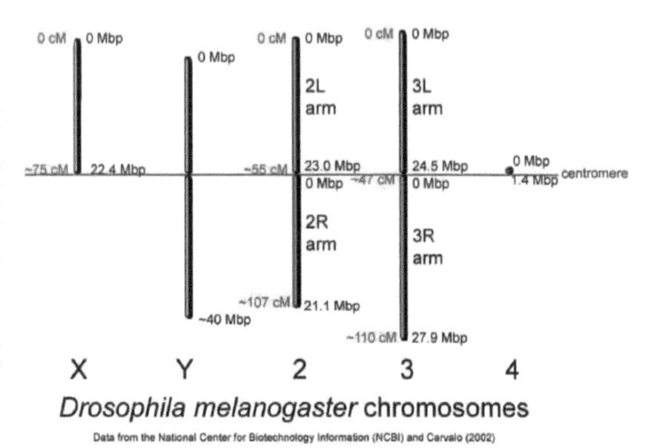

D. melanogaster chromosomes to scale with megabase-pair references oriented as in the National Center for Biotechnology Information database [17]. Centimorgan distances are approximate and estimated from the locations of selected mapped loci.

The genome of *D. melanogaster* (sequenced in 2000, and curated at the FlyBase database[15]) contains four pairs of chromosomes: an X/Y pair, and three autosomes labeled 2, 3, and 4. The fourth chromosome is so tiny that it is often ignored, aside from its important *eyeless* gene. The *D. melanogaster* sequenced genome of 165 million base pairs has been annotated[17] and contains approximately 13,767 protein-coding genes, which comprise ~20% of the genome out of a total of an estimated 14,000 genes. More than 60% of the genome appears to be functional non-protein-coding DNA[18] involved in gene expression control. Determination of sex in *Drosophila* occurs by the ratio of X chromosomes to autosomes, not because of the presence of a Y chromosome as in human sex determination. Although the Y chromosome is entirely heterochromatic, it contains at least 16 genes, many of which are thought to have male-related functions.[19]

Similarity to humans

About 75% of known human disease genes have a recognizable match in the genome of fruit flies,[20] and 50% of fly protein sequences have mammalian homologs. An online database called Homophila is available to search for human disease gene homologues in flies and vice versa.[21] *Drosophila* is being used as a genetic model for several human diseases including the neurodegenerative disorders Parkinson's, Huntington's, spinocerebellar ataxia and Alzheimer's disease. The fly is also being used to study mechanisms underlying aging and oxidative stress, immunity, diabetes, and cancer, as well as drug abuse.

Development

Embryogenesis in *Drosophila* has been extensively studied, as its small size, short generation time, and large brood size makes it ideal for genetic studies. It is also unique among model organisms in that cleavage occurs in a syncytium.

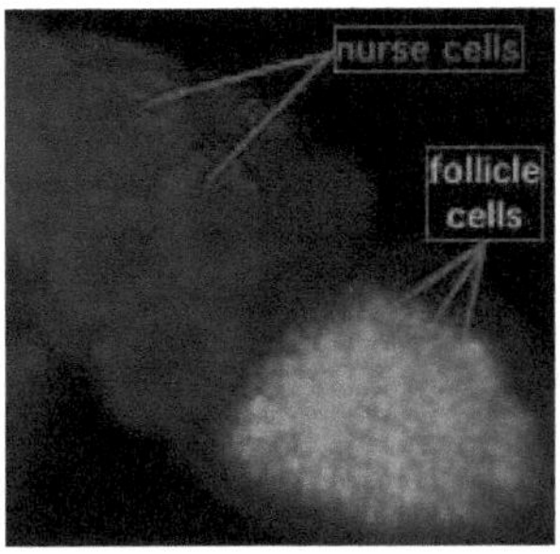

Drosophila melanogaster oogenesis

During oogenesis, cytoplasmic bridges called "ring canals" connect the forming oocyte to nurse cells. Nutrients and developmental control molecules move from the nurse cells into the oocyte. In the figure to the left, the forming oocyte can be seen to be covered by follicular support cells.

After fertilization of the oocyte the early embryo (or syncytial embryo) undergoes rapid DNA replication and 13 nuclear divisions until approximately 5000 to 6000 nuclei accumulate in the unseparated cytoplasm of the embryo. By the end of the 8th division most nuclei have migrated to the surface, surrounding the yolk sac (leaving behind only a few nuclei, which will become the yolk nuclei). After the 10th division the pole cells form at the posterior end of the embryo, segregating the germ line from the syncytium. Finally, after the 13th division cell membranes slowly invaginate, dividing the syncytium into individual somatic cells. Once this process is completed gastrulation starts.[22]

Nuclear division in the early *Drosophila* embryo happens so quickly there are no proper checkpoints so mistakes may be made in division of the DNA. To get around this problem, the nuclei that have made a mistake detach from their centrosomes and fall into the centre of the embryo (yolk sac), which will not form part of the fly.

The gene network (transcriptional and protein interactions) governing the early development of the fruit fly embryo is one of the best understood gene networks to date, especially the patterning along the antero-posterior (AP) and dorso-ventral (DV) axes (See under morphogenesis).[22]

The embryo undergoes well-characterized morphogenetic movements during gastrulation and early development, including germ-band extension, formation of several furrows, ventral invagination of the mesoderm, posterior and anterior invagination of endoderm (gut), as well as extensive body segmentation until finally hatching from the surrounding cuticle into a 1st-instar larva.

During larval development, tissues known as imaginal discs grow inside the larva. Imaginal discs develop to form most structures of the adult body, such as the head, legs, wings, thorax and genitalia. Cells of the imaginal disks are set aside during embryogenesis and continue to grow and divide during the larval stages—unlike most other cells of the larva, which have differentiated to perform specialized functions and grow without further cell division. At metamorphosis, the larva forms a pupa, inside which the larval tissues are reabsorbed and the imaginal tissues undergo extensive morphogenetic movements to form adult structures.

Immunity

Unlike mammals, Drosophila only have innate immunity and lack an adaptive immune response. The *D. melanogaster* immune system can be divided into two responses: humoral and cell-mediated. The former is a systemic response mediated through the Toll and imd pathways, which are parallel systems for detecting microbes. The Toll pathway in Drosophila is known as the homologue of Toll-Like pathways in mammals. Spatzle, a known ligand for the Toll pathway in flies, is produced in response to Gram-positive bacteria, parasites, and fungal infection. Upon infection, pro-Spatzle will be cleaved by protease SPE (Spatzle processing enzyme) to become active Spatzle, which then binds to the Toll receptor located on the cell surface (Fat body, hemocytes) and dimerise for activation of downstream NF-κB signaling pathways. On the other hand, the imd pathway is triggered by

Gram-negative bacteria through soluble and surface receptors (PGRP-LE and LC, respectively). *D. melanogaster* have a "fat body", which is thought to be homologous to the human liver. It is the primary secretory organ and produces antimicrobial peptides. These peptides are secreted into the hemolymph and bind infectious bacteria, killing them by forming pores in their cell walls. Years ago many drug companies wanted to purify these peptides and use them as antibiotics. Other than the fat body, hemocytes, the blood cells in drosophila, are known as the homologue of mammalian monocyte/macrophages, possessing a significant role in immune responses. It is known from the literature that in response to immune challenge, hemocytes are able to secret cytokines, for example Spatzle, to activate downstream signaling pathways in the fat body. However, the mechanism still remains unclear.

Behavioral genetics and neuroscience

In 1971, Ron Konopka and Seymour Benzer published "Clock mutants of *Drosophila melanogaster*", a paper describing the first mutations that affected an animal's behavior. Wild-type flies show an activity rhythm with a frequency of about a day (24 hours). They found mutants with faster and slower rhythms as well as broken rhythms—flies that move and rest in random spurts. Work over the following 30 years has shown that these mutations (and others like them) affect a group of genes and their products that comprise a biochemical or biological clock. This clock is found in a wide range of fly cells, but the clock-bearing cells that control activity are several dozen neurons in the fly's central brain.

Since then, Benzer and others have used behavioral screens to isolate genes involved in vision, olfaction, audition, learning/memory, courtship, pain and other processes, such as longevity.

The first learning and memory mutants (*dunce*, *rutabaga* etc.) were isolated by William "Chip" Quinn while in Benzer's lab, and were eventually shown to encode components of an intracellular signaling pathway involving cyclic AMP, protein kinase A and a transcription factor known as CREB. These molecules were shown to be also involved in synaptic plasticity in *Aplysia* and mammals.

Male flies sing to the females during courtship using their wing to generate sound, and some of the genetics of sexual behavior have been characterized. In particular, the fruitless gene has several different splice forms, and male flies expressing female splice forms have female-like behavior and vice-versa.

Furthermore, *Drosophila* has been used in neuropharmacological research, including studies of cocaine and alcohol consumption.

Vision

The compound eye of the fruit fly contains 760 unit eyes or ommatidia, and are one of the most advanced among insects. Each ommatidium contains 8 photoreceptor cells (R1-8), support cells, pigment cells, and a cornea. Wild-type flies have reddish pigment cells, which serve to absorb excess blue light so the fly isn't blinded by ambient light.

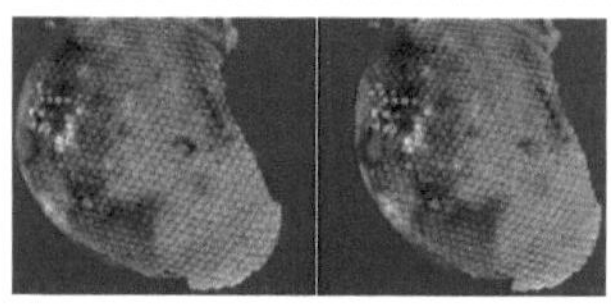

Stereo images of the fly eye

Each photoreceptor cell consists of two main sections, the cell body and the rhabdomere. The cell body contains the nucleus while the 100-μm-long rhabdomere is made up of toothbrush-like stacks of membrane called microvilli. Each microvillus is 1–2 μm in length and ~60 nm in diameter.[23] The membrane of the rhabdomere is packed with about 100 million rhodopsin molecules, the visual protein that absorbs light. The rest of the visual proteins are also tightly packed into the microvillar space, leaving little room for cytoplasm.

The photoreceptors in *Drosophila* express a variety of rhodopsin isoforms. The R1-R6 photoreceptor cells express Rhodopsin1 (Rh1), which absorbs blue light (480 nm). The R7 and R8 cells express a combination of either Rh3 or Rh4, which absorb UV light (345 nm and 375 nm), and Rh5 or Rh6, which absorb blue (437 nm) and green (508 nm) light respectively. Each rhodopsin molecule consists of an opsin protein covalently linked to a carotenoid chromophore, 11-cis-3-hydroxyretinal.[24]

As in vertebrate vision, visual transduction in invertebrates occurs via a G protein-coupled pathway. However, in vertebrates the G protein is transducin, while the G protein in invertebrates is Gq (dgq in *Drosophila*). When rhodopsin (Rh) absorbs a photon of light its chromophore, 11-cis-3-hydroxyretinal, is isomerized to all-trans-3-hydroxyretinal. Rh undergoes a conformational change into its active form, metarhodopsin. Metarhodopsin activates Gq, which in turn activates a phospholipase Cβ (PLCβ) known as NorpA.[25]

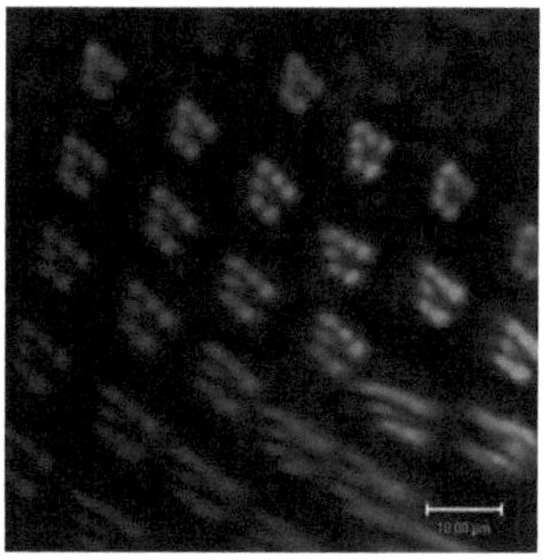

Expression of Rhodopsin1 (Rh1) in photoreceptors R1-R6

PLCβ hydrolyzes phosphatidylinositol (4,5)-bisphosphate (PIP_2), a phospholipid found in the cell membrane, into soluble inositol triphosphate (IP_3) and diacylgycerol (DAG), which stays in the cell membrane. DAG or a derivative of DAG causes a calcium selective ion channel known as TRP (transient receptor potential) to open and calcium and sodium flows into the cell. IP_3 is thought to bind to IP_3 receptors in the subrhabdomeric cisternae, an extension of the endoplasmic reticulum, and cause release of calcium, but this process doesn't seem to be essential for normal vision.[25]

Calcium binds to proteins such as calmodulin (CaM) and an eye-specific protein kinase C (PKC) known as InaC. These proteins interact with other proteins and have been shown to be necessary for shut off of the light response. In addition, proteins called arrestins bind metarhodopsin and prevent it from activating more Gq. A sodium-calcium exchanger known as CalX pumps the calcium out of the cell. It uses the inward sodium gradient to export calcium at a stoichiometry of 3 Na^+/ 1 Ca^{++}.[26]

TRP, InaC, and PLC form a signaling complex by binding a scaffolding protein called InaD. InaD contains five binding domains called PDZ domain proteins, which specifically bind the C termini of target proteins. Disruption of the complex by mutations in either the PDZ domains or the target proteins reduces the efficiency of signaling. For example, disruption of the interaction between InaC, the protein kinase C, and InaD results in a delay in inactivation of the light response.

Unlike vertebrate metarhodopsin, invertebrate metarhodopsin can be converted back into rhodopsin by absorbing a photon of orange light (580 nm).

Approximately two-thirds of the *Drosophila* brain is dedicated to visual processing.[27] Although the spatial resolution of their vision is significantly worse than that of humans, their temporal resolution is approximately ten times better.

Flight

The wings of a fly are capable of beating at up to 220 times per second. Flies fly via straight sequences of movement interspersed by rapid turns called saccades. During these turns, a fly is able to rotate 90 degrees in fewer than 50 milliseconds.

It was long thought that the characteristics of *Drosophila* flight were dominated by the viscosity of the air, rather than the inertia of the fly body. However, research in the lab of Michael Dickinson has indicated that flies perform banked turns, where the fly accelerates, slows down while turning, and accelerates again at the end of the turn. This indicates that inertia is the dominant force, as is the case with larger flying animals.[28] [29] Recent work, however, has shown that while the viscous effects on the insect body during flight may be negligible, the aerodynamic forces on the wings themselves actually cause fruit flies' turns to be damped viscously.[30]

See also

- Animal testing on invertebrates
- Genetically modified insect

References

[1] Meigen JW (1830) (in German) (PDF). *Systematische Beschreibung der bekannten europäischen zweiflügeligen Insekten. (Volume 6)* (https://dlib.stanford.edu:6521/text1/dd-ill/insekten6.pdf). Schulz-Wundermann. .

[2] Η Πύλη για την ελληνική γλώσσα (http://www.greek-language.gr/greekLang/index.html)

[3] Eric C. R. Reeve, ed (2001-06-23). "Drosophila melanogaster: The Fruit Fly" (http://books.google.com.au/books?id=JjLWYKqehRsC&pg=PA157&lpg=PA157&dq=drosophila+eggs+day+lifetime&source=bl&ots=V5BTOFQFeh&sig=JYiRArLjNyJy8SJylcncC26hh08&hl=en&ei=wTtLSpPOF4nENrTSwLgC&sa=X&oi=book_result&ct=result&resnum=2). *Encyclopedia of genetics*. USA: Fitzroy Dearborn Publishers, I. pp. 157. . Retrieved 2009-07-01.

[4] "FlyBase: A database of Drosophila genes and genomes" (http://flybase.bio.indiana.edu/). Genetics Society of America. 2009. . Retrieved August 11, 2009.

[5] Ashburner M, Thompson JN (1978). "The laboratory culture of *Drosophila*". In Ashburner M, Wright TRF. *The genetics and biology of* Drosophila. **2A**. Academic Press. 1–81.

[6] Ashburner M, Golic KG, Hawley RS (2005). *Drosophila: A Laboratory Handbook.* (2nd ed.). Cold Spring Harbor Laboratory Press. pp. 162–4. ISBN 0879697067.

[7] Bloomington *Drosophila* Stock Center (http://flystocks.bio.indiana.edu/) at Indiana University: Basic Methods of Culturing *Drosophila* (http://flystocks.bio.indiana.edu/Fly_Work/culturing.htm#stockkeeping)

[8] Chiang HC, Hodson AC (1950). "An analytical study of population growth in *Drosophila melanogaster*" (http://jstor.org/stable/1948580). *Ecological Monographs* **20** (3): 173–206. doi:10.2307/1948580. .

[9] Bakker K (1961). "An analysis of factors which determine success in competition for food among larvae of *Drosophila melanogaster*". *Archives Neerlandaises de Zoologie* **14**: 200–281.

[10] Pitnick S (1996). "Investment in testes and the cost of making long sperm in *Drosophila*". *American Naturalist* **148**: 57–80. doi:10.1086/285911.

[11] Houot, B.; Svetec, N.; Godoy-Herrera, R.; Ferveur, J. -F. (2010). "Effect of laboratory acclimation on the variation of reproduction-related characters in Drosophila melanogaster". *Journal of Experimental Biology* **213** (Pt 13): 2322. doi:10.1242/jeb.041566. PMID 20543131.

[12] Gilbert SF (2006). "9: Fertilization in Drosophila" (http://8e.devbio.com/article.php?ch=9&id=87). In 8th. *Developmental Biology*. Sinauer Associates. ISBN 978-0878932504. .

[13] Price C *et al.* (1999). "Sperm competition between Drosophila males involves both displacement and incapacitation". *Nature* **400** (6743): 449–452. Bibcode 1999Natur.400..449P. doi:10.1038/22755. PMID 10440373.

[14] Pierce, Benjamin A (2004). *Genetics: A Conceptual Approach* (2nd ed.). W. H. Freeman. ISBN 978-0716788812.

[15] Adams MD, Celniker SE, Holt RA, *et al.* (2000). "The genome sequence of *Drosophila melanogaster*" (http://www.sciencemag.org/cgi/content/abstract/287/5461/2185). *Science* **287** (5461): 2185–95. Bibcode 2000Sci...287.2185.. doi:10.1126/science.287.5461.2185. PMID 10731132. . Retrieved 2007-05-25.

[16] Azpiazu N, Frasch M (1993). "tinman and bagpipe: two homeo box genes that determine cell fates in the dorsal mesoderm of *Drosophila*". *Genes and Development* **7** (7b): 1325–1340. doi:10.1101/gad.7.7b.1325. PMID 8101173.

[17] Manning, Gerard (Oct 1, 2006). "Introduction to Drosophila" (http://www.ceolas.org/fly/intro.html). The WWW Virtual Library: Drosophila. p. 1. . Retrieved 2009-08-17.

[18] Halligan DL, Keightley PD (2006). "Ubiquitous selective constraints in the *Drosophila* genome revealed by a genome-wide interspecies comparison". *Genome Research* **16** (7): 875–84. doi:10.1101/gr.5022906. PMC 1484454. PMID 16751341.

[19] Carvalho, AB (2002). "Origin and evolution of the Drosophila Y chromosome". *Current Opinion in Genetics & Development* **12** (6852): 664–668. doi:10.1016/S0959-437X(02)00356-8.

[20] Reiter, LT; Potocki, L; Chien, S; Gribskov, M; Bier, E (2001). "A systematic analysis of human disease-associated gene sequences in *Drosophila melanogaster*". *Genome Research* **11** (6): 1114–1125. doi:10.1101/gr.169101. PMC 311089. PMID 11381037.

[21] Bier lab (2008). "Homophila: Human disease to Drosophila disease database" (http://superfly.ucsd.edu/homophila). University of California, San Diego. . Retrieved August 11, 2009.

[22] Katrin Weigmann, Robert Klapper, Thomas Strasser, Christof Rickert, Gerd Technau, Herbert Jäckle, Wilfried Janning & Christian Klämbt (2003). "FlyMove – a new way to look at development of *Drosophila*". *Trends in Genetics* **19** (6): 310–311. doi:10.1016/S0168-9525(03)00050-7. PMID 12801722.

[23] Hardie RC, Raghu P (2001). "Visual transduction in Drosophila". *Nature* **413** (6852): 186–93. doi:10.1038/35093002. PMID 11557987.

[24] Nichols R, Pak WL (1985). "Characterization of *Drosophila melanogaster* rhodopsin". *Journal of Biological Chemistry* **260** (23): 12670–4. PMID 3930500.

[25] Raghu P, Colley NJ, Webel R, *et al.* (2000). "Normal phototransduction in *Drosophila* photoreceptors lacking an InsP(3) receptor gene". *Molecular and Cellular Neuroscience* **15** (5): 429–45. doi:10.1006/mcne.2000.0846. PMID 10833300.

[26] Wang T, Xu H, Oberwinkler J, Gu Y, Hardie R, Montell C, *et al.* (2005). "Light activation, adaptation, and cell survival Functions of the Na+/Ca2+ exchanger CalX". *Neuron* **45** (3): 367–378. doi:10.1016/j.neuron.2004.12.046. PMID 15694299.
[27] Rein, K. and Zockler, M. and Mader, M.T. and Grubel, C. and Heisenberg, M. (2002). "The Drosophila Standard Brain". *Current Biology* **12** (3): 227–231. doi:10.1016/S0960-9822(02)00656-5. PMID 11839276.
[28] Caltech Press Release 4/17/2003 (http://mr.caltech.edu/media/Press_Releases/PR12371.html)
[29] S. Fry and M. Dickinson (2003). "The aerodynamics of free-flight maneuvers in *Drosophila*". *Science* **300** (5618): 495–8. Bibcode 2003Sci...300..495F. doi:10.1126/science.1081944. PMID 12702878.
[30] T. Hesselberg and F.-O. Lehmann (2007). "Turning behaviour depends on frictional damping in the fruit fly "*Drosophila*"". *The Journal of Experimental Biology* **210** (Pt 24): 4319–34. doi:10.1242/jeb.010389. PMID 18055621.

Further reading

- K. Haug-Collet, et al. (1999). "Cloning and characterization of a potassium-dependent sodium/calcium exchanger in *Drosophila*". *J. Cell Biol.* **147** (3): 659–70. doi:10.1083/jcb.147.3.659. PMC 2151195. PMID 10545508.
- R. Ranganathan, et al. (1995). "Signal transduction in *Drosophila* photoreceptors". *Annu. Rev. Neurosci.* **18**: 283–317. doi:10.1146/annurev.ne.18.030195.001435. PMID 7605064.
- Adams MD, *et al.* (2000). "The genome sequence of *Drosophila melanogaster*". *Science* **287** (5461): 2185–95. Bibcode 2000Sci...287.2185.. doi:10.1126/science.287.5461.2185. PMID 10731132.
- Kohler, Robert E. (1994). *Lords of the Fly:* Drosophila genetics *and the experimental life*. Chicago: University of Chicago Press. ISBN 0-226-45063-5.

Popular media

- "Inside the Fly Lab" (http://www.thirteen.org/curious/episodes/inside-the-fly-lab/) — broadcast by WGBH and PBS, in the program series "Curious", January 2008.
- "How a Fly Detects Poison" (http://whyfiles.org/shorties/285fly_taste/) — WhyFiles.org article describes how the fruit fly tastes a larva-killing chemical in food.

External links

External identifiers for *Drosophila melanogaster*	
NCBI Taxonomy	7227 (http://www.ncbi.nlm.nih.gov/Taxonomy/Browser/wwwtax.cgi?mode=Info&id=7227)
ITIS	146290 (http://www.itis.gov/servlet/SingleRpt/SingleRpt?search_topic=TSN&search_value=146290)
Encyclopedia of Life	733739 (http://www.eol.org/pages/733739)
Also found in: Wikispecies, ADW (http://animaldiversity.ummz.umich.edu/site/accounts/information/Drosophila_melanogaster.html)	

- A quick and simple introduction to *Drosophila melanogaster* (http://ceolas.org/VL/fly/intro.html)
- FlyBase — A Database of Drosophila Genes & Genomes (http://flybase.net/)
- NCBI page on *Drosophila melanogaster* (http://www.ncbi.nlm.nih.gov/mapview/map_search.cgi?taxid=7227)
- The WWW Virtual Library: *Drosophila* (http://www.ceolas.org/VL/fly/)
- The Berkeley *Drosophila* Genome Project (http://www.fruitfly.org/)
- FlyMove (http://flymove.uni-muenster.de/)
- The Interactive Fly — A guide to *Drosophila* genes and their roles in development (http://www.sdbonline.org/fly/aimain/1aahome.htm)
- *Drosophila* Nomenclature — naming of genes (http://www.flynome.com/index.html)
- Make Your Own Fruit Fly Trap (http://lancaster.unl.edu/pest/resources/FruitFlyTrap.shtml)

- Illustrates a simple to make non-toxic Vinegar fly trap (http://ohioline.osu.edu/hyg-fact/2000/2109.html)
- Measurement of Courtship Behavior in Drosophila melanogaster (http://www.cshprotocols.org/cgi/content/full/2007/20/pdb.prot4847)
- Maintenance of a Drosophila Laboratory: General Procedures (http://www.cshprotocols.org/cgi/content/full/2007/6/pdb.ip35)
- Transcript In Situ Hybridization of Whole-Mount Embryos for Phenotype Analysis of RNAi-Treated Drosophila (http://www.cshprotocols.org/cgi/content/full/2006/21/pdb.prot4519)
- Injection of dsRNA into Drosophila Embryos for RNA Interference (RNAi) (http://www.cshprotocols.org/cgi/content/full/2008/3/pdb.prot4918)

Louse

Phthiraptera	

Light micrograph of *Fahrenholzia pinnata*	
Scientific classification	
Kingdom:	Animalia
Phylum:	Arthropoda
Class:	Insecta
Subclass:	Pterygota
Infraclass:	Neoptera
Order:	**Phthiraptera** Haeckel, 1896
Suborders	
Anoplura Rhyncophthirina Ischnocera Amblycera	

Lice (singular: **louse**) is the common name for over 3000 species of wingless insects of the order Phthiraptera; three of which are classified as human disease agents. They are obligate ectoparasites of every avian and mammalian order except for monotremes (the platypus and echidnas), bats, whales, dolphins, porpoises and pangolins.

Biology

Most lice are scavengers, feeding on skin and other debris found on the host's body, but some species feed on sebaceous secretions and blood. Most are found only on specific types of animal, and, in some cases, only to a particular part of the body; some animals are known to host up to fifteen different species, although one to three is typical for mammals, and two to six for birds. For example, in humans, different species of louse inhabit the scalp and pubic hair. Lice generally cannot survive for long if removed from their host.[1]

A louse's color varies from pale beige to dark gray; however, if feeding on blood, it may become considerably darker. Female lice are usually more common than the males, and some species are even known to be parthenogenetic. A louse's egg is commonly called a nit. Many lice attach their eggs to their host's hair with specialized saliva; the saliva/hair bond is very difficult to sever without specialized products. Lice inhabiting birds, however, may simply leave their eggs in parts of the body inaccessible to preening, such as the interior of feather shafts. Living lice eggs tend to be pale white. Dead lice eggs are more yellow.[1]

Lice are exopterygotes, being born as miniature versions of the adult, known as nymphs. The young moult three times before reaching the final adult form, which they usually reach within a month of hatching.[1]

Ecology

Lice are optimal model organisms to study the ecology of contagious pathogens since their quantities, sex-ratios etc. are easier to quantify than those of other pathogens. The ecology of avian lice have been studied more intensively than that of mammal lice.

A few major trends

1. The average number of lice per host tends to be higher in large-bodied bird species than in small ones.[2]
2. Louse individuals exhibit an aggregated distribution across bird individuals, i.e. most lice live on a few bird, while most birds are relatively free of lice. This pattern is more pronounced in territorial than in colonial – more social – bird species.[3]
3. Host taxa that dive under the water surface to feed on aquatic prey harbour fewer taxa of lice.[4] [5]
4. Bird taxa that are capable to exert stronger antiparasitic defense – such as stronger T cell immune response or larger uropygial glands – harbour more taxa of Amblyceran lice than others.[6] [7]
5. Temporal bottlenecks in host population size may couse a long-lasting reduction of louse taxonomic richness.[8] E.g., birds introduced into New Zealand host fewer species of lice there than in Europe.[9] [10]
6. Louse sex ratios are more balanced in more social hosts and more female-biased in less social hosts, presumably due to the stronger isolation among louse subpopulations (living on separate birds) in the latter case.[11]

A few effects of lice infestation upon the host

1. Lice may reduce host life expectancy.[12]
2. Lice may transmit microbial diseases or helminth parasites.[13]
3. Ischnoceran lice may reduce the thermoregulation effect of the plumage, thus heavily infected birds loss more heat than other ones.[14]
4. Lice infestation is a disadvantage in the context of sexual rivalry.[15] [16]

Classification

The order has traditionally been divided into two suborders, the sucking lice (Anoplura) and the chewing lice (Mallophaga); however, recent classifications suggest that the Mallophaga are paraphyletic and four suborders are now recognised:

- Anoplura: sucking lice, occurring on mammals exclusively
- Rhyncophthirina: parasites of elephants and warthogs
- Ischnocera: mostly avian chewing lice, however, one family parasitize mammals
- Amblycera: a primitive suborder of chewing lice, widspread on birds, however, also live on South-American And Australian mammals

It has been suggested that the order is contained by the Troctomorpha suborder of Psocoptera.

Lice in humans

Humans host three different kinds of lice: head lice, body lice, and pubic lice. Lice infestations can be controlled with lice combs, and medicated shampoos or washes. Adult and nymphal lice can survive on sheep-shearers' moccasins for up to 10 days, but microwaving the footwear for five minutes in a plastic bag will kill the lice.[17]

Human lice and DNA discoveries

Lice have been the subject of significant DNA research that has led to discoveries on human evolution. For example, recent DNA evidence suggests that pubic lice spread to humans approximately 2,000,000 years ago from gorillas.[18] Additionally, the DNA differences between head lice and body lice provide corroborating evidence that humans started losing body hair, also about 2,000,000 years ago.[19]

Gallery

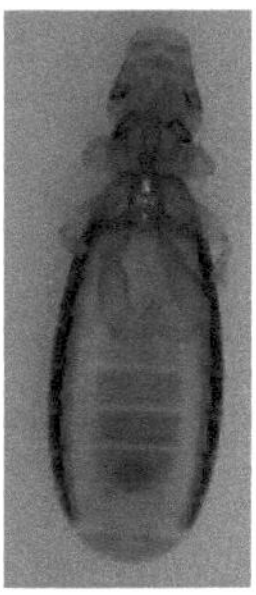

Ricinus bombycillae, an Amblyceran louse from the bohemian waxwing

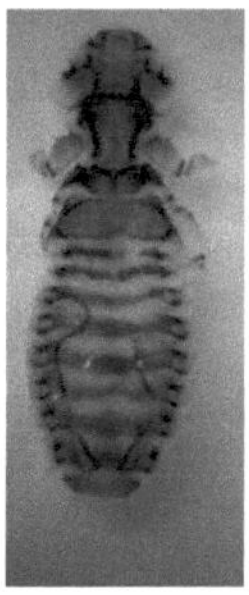

Trinoton anserinum, an Amblyceran louse from a mute swan.

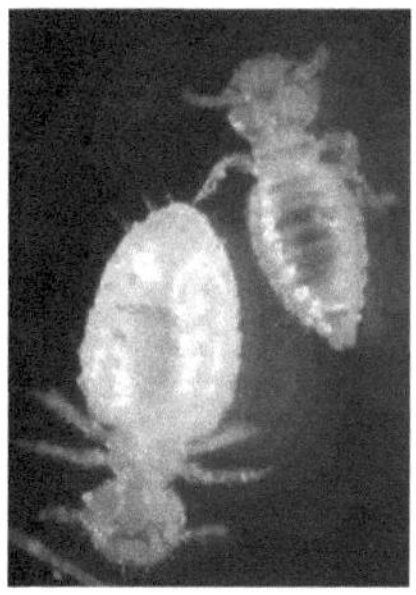

Damalinia limbata is an Ischnoceran louse from goats. The male is smaller than the female.

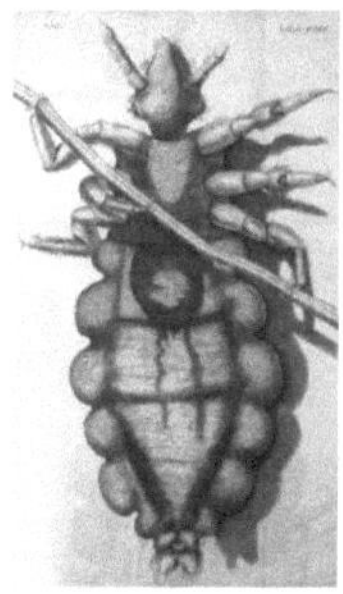

Diagram of a louse, by Robert Hooke, 1667.

See also

- Pest (organism)
- Use of DNA in forensic entomology
- Whale louse, unrelated crustaceans that live parasitically on marine mammals

References

[1] H. V. Hoell, J. T. Doyen & A. H. Purcell (1998). *Introduction to Insect Biology and Diversity* (2nd ed.). Oxford University Press. pp. 407–409. ISBN 0-19-510033-6.

[2] Rózsa L 1997. Patterns in the abundance of avian lice (Phthiraptera: Amblycera, Ischnocera) (http://www.zoologia.hu/list/patterns in the abundance.pdf). *Journal of Avian Biology*, **28,** 249–254.

[3] Rékási J et al. 1997. Patterns in the distribution of avian lice (Phthiraptera: Amblycera, Ischnocera) (http://www.zoologia.hu/list/patterns in the distribution.pdf). *Journal of Avian Biology*, **28,** 150–156.

[4] Felső B et al. 2006. Reduced taxonomic richness of lice (Insecta: Phthiraptera) in diving birds. (http://www.zoologia.hu/list/felso_rozsa.pdf) *Journal of Parasitology*, **92,** 867–869.

[5] Felső B et al. 2007. Diving behaviour reduces genera richness of lice (Insecta: Phthiraptera) of mammals (http://www.zoologia.hu/list/felso_rozsa2.pdf). *Acta Parasitologica*, **52,** 82–85.

[6] Møller AP et al. 2005. Parasite biodiversity and host defenses: Chewing lice and immune response of their avian hosts (http://www.zoologia.hu/list/Moller_Rozsa_Oecologia.pdf). *Oecologia*, **142,** 169–176.

[7] Møller AP et al. 2010. Ectoparasites, uropygial glands and hatching success in birds (http://www.zoologia.hu/list/Moller et al 2010.pdf). *Oecologia*, **163**, 303–311.

[8] Rózsa L 1993. Speciation patterns of ectoparasites and "straggling" lice (http://www.zoologia.hu/list/straggling.pdf). *International Journal for Parasitology*, **23**, 859–864.

[9] Paterson AM et al. 1999. How Frequently Do Avian Lice Miss the Boat? Implications for Coevolutionary Studies (http://sysbio.oxfordjournals.org/cgi/reprint/48/1/214.pdf) *Systematic Biology*, **48**, 214–223

[10] MacLeod C et al. 2010. Parasites lost – do invaders miss the boat or drown on arrival? (http://onlinelibrary.wiley.com/doi/10.1111/j.1461-0248.2010.01446.x/abstract) *Ecology Letters*,

[11] Rózsa L et al. 1996. Relationship of host coloniality to the population ecology of avian lice (Insecta: Phthiraptera) (http://www.zoologia.hu/list/relationship of host coloniality.pdf). *Journal of Animal Ecology*, **65**, 242–248.

[12] Brown CR et al. 1995. Ectoparasites reduce long-term surviviorship of their avian host (http://www.rannala.org/reprints/1995/Brown1995a.pdf). *Proceedings of the Royal Society of London B* **262**, 313–319.

[13] Barlett CM 1993. Lice (Amblycera and Ischnocera) as vectors of Eulimdana spp. (Nematoda: Filarioidea) in Charadriiform birds and the necessity of short reproductive periods in adult worms (http://www.jstor.org/pss/3283282). *Journal of Parasitol*, **79**, 85–91.

[14] Booth DT et al. 1993. Experimental demonstration of the energetic cost of parasitism in free-ranging hosts (http://darwin.biology.utah.edu/PubsHTML/PDF-Files/18.pdf). *Proceedings of the Royal Society of London B*, **253**, 125–129.

[15] Clayton DH 1990. Mate choice in experimentally parasitized rock doves: lousy males lose (http://darwin.biology.utah.edu/PubsHTML/PDF-Files/11.pdf). *American Zoologist*, **30**, 251–262.

[16] Garamszegi LZ et al. 2005. Age-dependent health status and song characteristics (http://beheco.oxfordjournals.org/cgi/content/full/16/3/580). Beahvioral Ecology, ***16*****,580–591.**

[17] "Sheep parasites" (http://www2.dpi.qld.gov.au/sheep/10042.html). Government of Queensland. . Retrieved November 10, 2008.

[18] David L. Reed, Jessica E. Light, Julie M. Allen & Jeremy J. Kirchman (2007). "Pair of lice lost or parasites regained: the evolutionary history of anthropoid primate lice" (http://www.biomedcentral.com/1741-7007/5/7). *BMC Biology* **5**: 7. doi:10.1186/1741-7007-5-7. PMC 1828715. PMID 17343749. .

[19] John Travis (August 23, 2003). *The naked truth? Lice hint at a recent origin of clothing* (http://findarticles.com/p/articles/mi_m1200/is_8_164/ai_107897267/). **164**. Science News. pp. 118. .

External links

- National Pesticide Information Center – Understanding and Controlling Lice (http://npic.orst.edu/pest/lice.html)
- Bed-hopping led humans to 3 million-year itch (http://www.theaustralian.news.com.au/story/0,20867,21346795-2703,00.html)
- Extensive scientific information at Phthiraptera Central (http://www.phthiraptera.org)
- body and head lice (http://entomology.ifas.ufl.edu/creatures/urban/human_lice.htm) on the University of Florida/Institute of Food and Agricultural Sciences Featured Creatures Web site
- crab louse (http://entomology.ifas.ufl.edu/creatures/urban/crab_louse.htm) on the University of Florida/Institute of Food and Agricultural Sciences Featured Creatures Web site
- WebMD Lice Info (http://children.webmd.com/tc/lice-topic-overview)
- Lice (http://entomology.ifas.ufl.edu/fasulo/vector/chapter_04.htm) chapter in United States Environmental Protection Agency and University of Florida/Institute of Food and Agricultural Sciences National Public Health Pesticide Applicator Training Manual
- *Pediculus humanus capitis* head louse facts, myths, life cycle at MetaPathogen (http://www.metapathogen.com/lice/phumanus/)

Mold

Molds (or **moulds**; see spelling differences) are fungi that grow in the form of multicellular filaments called hyphae.[1] Molds are considered to be microbes but microscopic fungi that grow as single cells are called yeasts. A connected network of these tubular branching hyphae has multiple, genetically identical nuclei and is considered a single organism, referred to as a colony.

Mold growing on a clementine

Molds do not form a specific taxonomic or phylogenetic grouping, but can be found in the divisions Zygomycota, Deuteromycota and Ascomycota. Some molds cause disease or food spoilage, others play an important role in biodegradation or in the production of various foods, beverages, antibiotics and enzymes.

Biology

There are thousands of known species of molds which include opportunistic pathogens, saprotrophs, aquatic species, and thermophiles.[2] Like all fungi, molds derive energy not through photosynthesis but from the organic matter in which they live. Typically, molds secrete hydrolytic enzymes, mainly from the hyphal tips. These enzymes degrade complex biopolymers such as starch, cellulose and lignin into simpler substances which can be absorbed by the hyphae. In this way, molds play a major role in causing decomposition of organic material, enabling the recycling of nutrients throughout ecosystems. Many molds also secrete mycotoxins which, together with hydrolytic enzymes, inhibit the growth of competing microorganisms.

Molds reproduce through small spores,[2] which may contain a single nucleus or be multinucleate. Mold spores can be asexual (the products of mitosis) or sexual (the products of meiosis); many species can produce both types. Mold spores may remain airborne indefinitely, may cling to clothing or fur, or may be able to survive extremes of temperature and pressure.

Although molds grow on dead organic matter everywhere in nature, their presence is only visible to the unaided eye when mold colonies grow. A mold colony does not comprise discrete organisms, but an interconnected network of hyphae called a mycelium. Nutrients and in some cases organelles may be transported throughout the mycelium. In artificial environments like buildings, humidity and temperature are often stable enough to foster the growth of mold colonies, commonly seen as a downy or furry coating growing on food or other surfaces.

Many molds can begin growing at 4 °C (39 °F), the temperature within a typical refrigerator, or less. When conditions do not enable growth, molds may remain alive in a dormant state depending on the species, within a large range of temperatures before they die. The many different mold species vary enormously in their tolerance to temperature and humidity extremes. Certain molds can survive harsh conditions such as the snow-covered soils of Antarctica, refrigeration, highly acidic solvents, anti-bacterial soap and even petroleum products such as jet fuel.

Xerophilic molds use the humidity in the air as their only water source; other molds need more moisture.

Common molds

- Acremonium
- Aspergillus
- Cladosporium
- Fusarium
- Mucor
- Penicillium
- Rhizopus
- Stachybotrys
- Trichoderma
- Alternaria

Food production

The Kōji () molds are a group of *Aspergillus* species, notably *Aspergillus oryzae*, and secondarily *A. sojae*, that have been cultured in eastern Asia for many centuries. They are used to ferment a soybean and wheat mixture to make soybean paste and soy sauce. *Koji* molds break down the starch in rice, barley, sweet potatoes, etc, a process called saccharification, in the production of *sake*, *shōchū* and other distilled spirits. *Koji* molds are also used in the preparation of Katsuobushi.

Red rice yeast is a product of the mold *Monascus purpureus* grown on rice, and is common in Asian diets. The yeast contains several compounds collectively known as monacolins, which are known to inhibit cholesterol synthesis.[3] According to a study published in the journal *Mayo Clinic Proceedings* by Dr. David Becker, red rice yeast used as a dietary supplement, combined with fish oil and healthy lifestyle changes, may help reduce "bad" cholesterol as effectively as certain commercial statin drugs.[4]

Some sausages, such as salami, use starter cultures[5] in their production, to improve flavour and reduce spoilage during curing.

Other molds that have been used in food production include:

- *Fusarium venenatum* – quorn
- *Geotrichum candidum* – cheese
- *Neurospora sitophila* – oncom
- *Penicillium* spp. – cheese
- *Rhizomucor miehei* – rennet for making vegetarian and other cheese
- *Rhizopus oligosporus* – tempeh
- *Ustilago maydis* – filling in tortilla-based foods

Drug creation

Alexander Fleming's famous discovery of the antibiotic penicillin involved the mold *Penicillium chrysogenum*.

Several cholesterol-lowering drugs (such as Lovastatin, from *Aspergillus terreus*) are derived from molds.

The immunosuppressant drug cyclosporine, used to suppress the rejection of transplanted organs, is derived from the mold *Tolypocladium inflatum*.

Health effects

Molds are ubiquitous in nature, and mold spores are a common component of household and workplace dust. However, when mold spores are present in large quantities, they can present a health hazard to humans, potentially causing allergic reactions and respiratory problems.

Some molds also produce mycotoxins that can pose serious health risks to humans and animals. Some studies claim that exposure to high levels of mycotoxins can lead to neurological problems and in some cases death. Prolonged exposure, e.g. daily workplace exposure, may be particularly harmful. Research on the health effects of mold has not been conclusive. The term "toxic mold" refers to molds that produce mycotoxins, such as *Stachybotrys chartarum*, and not to all molds in general.[6]

Mold in the home can usually be found in damp, dark or steam filled areas e.g. bathroom or kitchen, cluttered storage areas, recently flooded areas, basement areas, plumbing spaces, areas with poor ventilation and outdoors in humid environments. Symptoms caused by mold allergy are watery, itchy eyes, a chronic cough, headaches or migraines, difficulty breathing, rashes, tiredness, sinus problems, nasal blockage and frequent sneezing.

Growth in buildings and homes

Mold growth in buildings can lead to a variety of health issues. Various practices can be followed to mitigate mold issues in buildings, the most important of which is to reduce moisture levels that can facilitate mold growth.[6] Removal of affected materials after the source of moisture has been reduced and/or eliminated may be necessary for remediation.

Gallery

Bread mold is one of the most common types of mold, and can cover a loaf of bread in less than three days.

Mold covering a decaying peach over a period of six days. The frames were taken approximately 12 hours apart. There are 12 frames of changes.

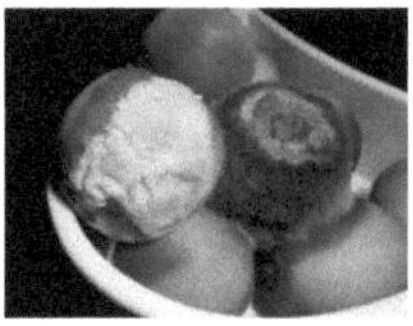

Moldy nectarines that were in a refrigerator. The nectarine with black mold is also affecting the nectarine underneath.

The mold that attacks bread is extremely common, and can cover a loaf almost completely within three days.

A bowl of moldy strawberries after being left in a room for several days.

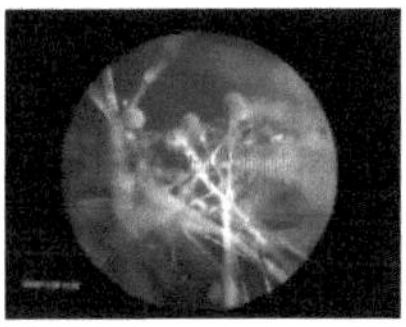

Mold cultured from dust shaken from an automotive cabin air filter. 10× objective, 15× eyepiece; each numbered tick is 122 μm.

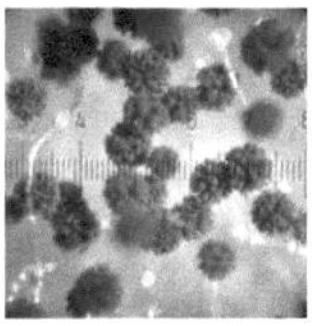

Mold grown from a nasal mucus sample. 10× objective, 15× eyepiece; each numbered tick is 122 μm.

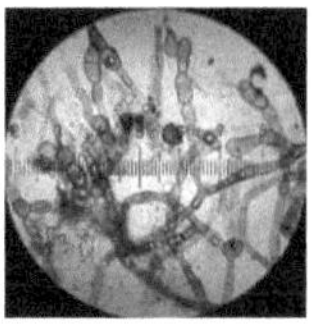

Mildew from a wooden soap holder. Numbered ticks are 11 μm apart.

Mildew growing on a plastic shower curtain. 10× objective, 15× eyepiece; each numbered tick is 122 μm.

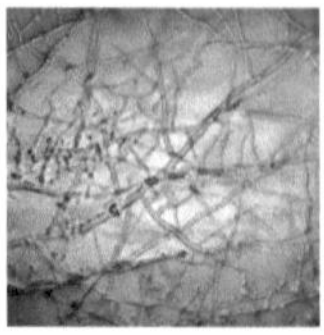

Mycelium of unidentified mold. This image covers a one-millimeter square.

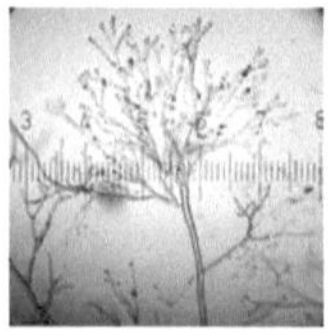

Another mycelium, of another mold. Numbered ticks are 230 μm apart.

Stilton cheese contains edible mold.

A moldy tomato

Moldy bread

Detailed picture of mold on tomato

Moldy coffee

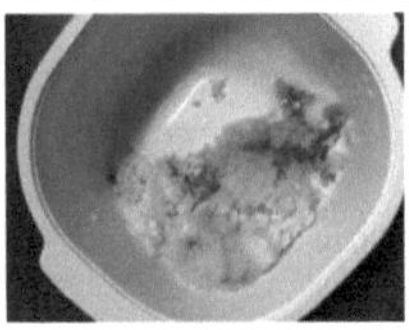

Mold on rice

See also

- Bioaerosol
- Decay
- Medicinal mushrooms
- Mildew
- Mold mite
- Mycorrhiza
- Oomycete

References

[1] Madigan M; Martinko J (editors). (2005). *Brock Biology of Microorganisms* (11th ed.). Prentice Hall. ISBN 0131443291. OCLC 57001814.

[2] Ryan KJ; Ray CG (editors) (2004). *Sherris Medical Microbiology* (4th ed.). McGraw Hill. pp. 633–8. ISBN 0838585299.

[3] "Red yeast rice (Monascus purpureus)" (http://www.mayoclinic.com/health/red-yeast-rice/NS_patient-redyeast). Mayo Clinic. 2009-09-01. . Retrieved 2010-02-01.

[4] "Study: Red Rice Yeast Helps Cut Bad Cholesterol" (http://www.npr.org/templates/story/story.php?storyId=92103272). National Public Radio. 2008-07-01. . Retrieved 2010-02-01.

[5] L. H. Stahnke, L. O. Sunesen (2003-11). "Mould starter cultures for dry sausages—selection, application and effects" (http://www.sciencedirect.com/science?_ob=ArticleURL&_udi=B6T9G-48V817P-1&_user=10&_rdoc=1&_fmt=&_orig=search&_sort=d&view=c&_acct=C000050221&_version=1&_urlVersion=0&_userid=10&md5=d49cf5f0622cb4266ceba3c7f36fa924). *Meat Science* **65** (3): 935–948. doi:10.1016/S0309-1740(02)00281-4. . Retrieved 2008-06-06.

[6] Indoor Environmental Quality: Dampness and Mold in Buildings (http://www.cdc.gov/niosh/topics/indoorenv/mold.html). National Institute for Occupational Safety and Health. August 1, 2008.

External links

- The EPA's guide to mold (http://www.epa.gov/mold/moldguide.html)

Fungus

Fungi
Temporal range: Early Devonian–Recent (but see text)

Clockwise from top left: *Amanita muscaria*, a basidiomycete; *Sarcoscypha coccinea*, an ascomycete; bread covered in mold; a chytrid; a *Penicillium* conidiophore.

Scientific classification	
Domain:	Eukaryota
Unranked:	Opisthokonta
Kingdom:	**Fungi** (L., 1753) R.T. Moore, 1980[1]

Subkingdoms/Phyla/Subphyla[2]

- Blastocladiomycota
- Chytridiomycota
- Glomeromycota
- Microsporidia
- Neocallimastigomycota
- Dikarya (inc. Deuteromycota)
  - Ascomycota
    - Pezizomycotina
    - Saccharomycotina
    - Taphrinomycotina
  - Basidiomycota
    - Agaricomycotina
    - Pucciniomycotina
    - Ustilaginomycotina
- Subphyla Incertae sedis
  - Entomophthoromycotina
  - Kickxellomycotina
  - Mucoromycotina
  - Zoopagomycotina

A **fungus** (pronounced /ˈfʌŋɡəs/; pl. fungi[3] or funguses[4]) is a member of a large group of eukaryotic organisms that includes microorganisms such as yeasts and molds (British English: moulds), as well as the more familiar mushrooms. These organisms are classified as a kingdom, **Fungi**, which is separate from plants, animals, and bacteria. One major difference is that fungal cells have cell walls that contain chitin, unlike the cell walls of plants, which contain cellulose. These and other differences show that the fungi form a single group of related organisms, named the *Eumycota* (*true fungi* or *Eumycetes*), that share a common ancestor (a *monophyletic group*). This fungal group is distinct from the structurally similar myxomycetes (slime molds) and oomycetes (water molds). The discipline of biology devoted to the study of fungi is known as mycology, which is often regarded as a branch of botany, even though genetic studies have shown that fungi are more closely related to animals than to plants.

Abundant worldwide, most fungi are inconspicuous because of the small size of their structures, and their cryptic lifestyles in soil, on dead matter, and as symbionts of plants, animals, or other fungi. They may become noticeable when fruiting, either as mushrooms or molds. Fungi perform an essential role in the decomposition of organic matter and have fundamental roles in nutrient cycling and exchange. They have long been used as a direct source of food, such as mushrooms and truffles, as a leavening agent for bread, and in fermentation of various food products, such as wine, beer, and soy sauce. Since the 1940s, fungi have been used for the production of antibiotics, and, more recently, various enzymes produced by fungi are used industrially and in detergents. Fungi are also used as biological pesticides to control weeds, plant diseases and insect pests. Many species produce bioactive compounds called mycotoxins, such as alkaloids and polyketides, that are toxic to animals including humans. The fruiting structures of a few species contain psychotropic compounds and are consumed recreationally or in traditional spiritual ceremonies. Fungi can break down manufactured materials and buildings, and become significant pathogens of humans and other animals. Losses of crops due to fungal diseases (e.g. rice blast disease) or food spoilage can have a large impact on human food supplies and local economies.

The fungus kingdom encompasses an enormous diversity of taxa with varied ecologies, life cycle strategies, and morphologies ranging from single-celled aquatic chytrids to large mushrooms. However, little is known of the true biodiversity of Kingdom Fungi, which has been estimated at around 1.5 million species, with about 5% of these having been formally classified. Ever since the pioneering 18th and 19th century taxonomical works of Carl Linnaeus, Christian Hendrik Persoon, and Elias Magnus Fries, fungi have been classified according to their morphology (e.g., characteristics such as spore color or microscopic features) or physiology. Advances in molecular genetics have opened the way for DNA analysis to be incorporated into taxonomy, which has sometimes challenged the historical groupings based on morphology and other traits. Phylogenetic studies published in the last decade have helped reshape the classification of Kingdom Fungi, which is divided into one subkingdom, seven phyla, and ten subphyla.

Etymology

The English word *fungus* is directly adopted from the Latin *fungus* (mushroom), used in the writings of Horace and Pliny.[5] This in turn is derived from the Greek word *sphongos*/σφογγος ("sponge"), which refers to the macroscopic structures and morphology of mushrooms and molds; the root is also used in other languages, such as the German *Schwamm* ("sponge"), *Schimmel* ("mold"), and the French *champignon* and the Spanish *champiñon* (which both mean "mushroom").[6] The use of the word *mycology*, which is derived from the Greek *mykes*/μύκης (mushroom) and *logos*/λόγος (discourse),[7] to denote the scientific study of fungi is thought to have originated in 1836 with English naturalist Miles Joseph Berkeley's publication *The English Flora of Sir James Edward Smith, Vol. 5.*[6]

Characteristics

Before the introduction of molecular methods for phylogenetic analysis, taxonomists considered fungi to be members of the Plant Kingdom because of similarities in lifestyle: both fungi and plants are mainly immobile, and have similarities in general morphology and growth habitat. Like plants, fungi often grow in soil, and in the case of mushrooms form conspicuous fruiting bodies, which sometimes bear resemblance to plants such as mosses. The fungi are now considered a separate kingdom, distinct from both plants and animals, from which they appear to have diverged around one billion years ago.[8] [9] Some morphological, biochemical, and genetic features are shared with other organisms, while others are unique to the fungi, clearly separating them from the other kingdoms:

Shared features:

- With other eukaryotes: As other eukaryotes, fungal cells contain membrane-bound nuclei with chromosomes that contain DNA with noncoding regions called introns and coding regions called exons. In addition, fungi possess membrane-bound cytoplasmic organelles such as mitochondria, sterol-containing membranes, and ribosomes of the 80S type.[10] They have a characteristic range of soluble carbohydrates and storage compounds, including sugar alcohols (e.g., mannitol), disaccharides, (e.g., trehalose), and polysaccharides (e.g., glycogen, which is also found in animals[11]).
- With animals: Fungi lack chloroplasts and are heterotrophic organisms, requiring preformed organic compounds as energy sources.[12]
- With plants: Fungi possess a cell wall[13] and vacuoles.[14] They reproduce by both sexual and asexual means, and like basal plant groups (such as ferns and mosses) produce spores. Similar to mosses and algae, fungi typically have haploid nuclei.[15]
- With euglenoids and bacteria: Higher fungi, euglenoids, and some bacteria produce the amino acid L-lysine in specific biosynthesis steps, called the α-aminoadipate pathway.[16] [17]
- The cells of most fungi grow as tubular, elongated, and thread-like (filamentous) structures and are called hyphae, which may contain multiple nuclei and extend at their tips. Each tip contains a set of aggregated vesicles—cellular structures consisting of proteins, lipids, and other organic molecules—called Spitzenkörper.[18] Both fungi and oomycetes grow as filamentous hyphal cells.[19] In contrast, similar-looking organisms, such as filamentous green algae, grow by repeated cell division within a chain of cells.[11]
- In common with some plant and animal species, more than 60 fungal species display the phenomenon of bioluminescence.[20]

Unique features:

- Some species grow as single-celled yeasts that reproduce by budding or binary fission. Dimorphic fungi can switch between a yeast phase and a hyphal phase in response to environmental conditions.[21]
- The fungal cell wall is composed of glucans and chitin; while the former compounds are also found in plants and the latter in the exoskeleton of arthropods,[22] [23] fungi are the only organisms that combine these two structural molecules in their cell wall. In contrast to plants and the oomycetes, fungal cell walls do not contain cellulose.[24]

Most fungi lack an efficient system for long-distance transport of water and nutrients, such as the xylem and phloem in many plants. To overcome these limitations, some fungi, such as *Armillaria*, form rhizomorphs,[25] that resemble and perform functions similar to the roots of plants. Another characteristic shared with plants includes a biosynthetic pathway for producing terpenes that uses mevalonic acid and pyrophosphate as chemical building blocks.[26] However, plants have an additional terpene pathway in their chloroplasts, a structure fungi do not possess.[27] Fungi produce several secondary metabolites that are similar or identical in structure to those made by plants.[26] Many of the plant and fungal enzymes that make these compounds differ from each other in sequence and other characteristics, which indicates separate origins and evolution of these enzymes in the fungi and plants.[26] [28]

Omphalotus nidiformis, a bioluminescent mushroom

Diversity

Fungi have a worldwide distribution, and grow in a wide range of habitats, including extreme environments such as deserts or areas with high salt concentrations[29] or ionizing radiation,[30] as well as in deep sea sediments.[31] Some can survive the intense UV and cosmic radiation encountered during space travel.[32] Most grow in terrestrial environments, though several species live partly or solely in aquatic habitats, such as the chytrid fungus *Batrachochytrium dendrobatidis*, a parasite that has been responsible for a worldwide decline in amphibian populations. This organism spends part of its life cycle as a motile zoospore, enabling it to propel itself through water and enter its amphibian host.[33] Other examples of aquatic fungi include those living in hydrothermal areas of the ocean.[34]

Around 100,000 species of fungi have been formally described by taxonomists,[35] but the global biodiversity of the fungus kingdom is not fully understood.[36] On the basis of observations of the ratio of the number of fungal species to the number of plant species in selected environments, the fungal kingdom has been estimated to contain about 1.5 million species;[37] a recent (2011) estimate suggests there may be over 5 million species.[38] In mycology, species have historically been distinguished by a variety of methods and concepts. Classification based on morphological characteristics, such as the size and shape of spores or fruiting structures, has traditionally dominated fungal taxonomy.[39] Species may also be distinguished by their biochemical and physiological characteristics, such as their ability to metabolize certain biochemicals, or their reaction to chemical tests. The biological species concept discriminates species based on their ability to mate. The application of molecular tools, such as DNA sequencing and phylogenetic analysis, to study diversity has greatly enhanced the resolution and added robustness to estimates of genetic diversity within various taxonomic groups.[40]

Morphology

Microscopic structures

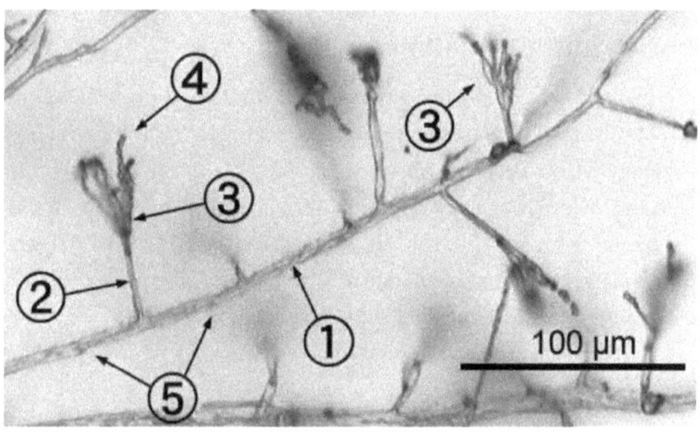

An environmental isolate of *Penicillium*
1. hypha 2. conidiophore 3. phialide 4. conidia 5. septa

Most fungi grow as hyphae, which are cylindrical, thread-like structures 2–10 µm in diameter and up to several centimeters in length. Hyphae grow at their tips (apices); new hyphae are typically formed by emergence of new tips along existing hyphae by a process called *branching*, or occasionally growing hyphal tips bifurcate (fork) giving rise to two parallel-growing hyphae.[41] The combination of apical growth and branching/forking leads to the development of a mycelium, an interconnected network of hyphae.[21] Hyphae can be either septate or coenocytic: septate hyphae are divided into compartments separated by cross walls (internal cell walls, called septa, that are formed at right angles to the cell wall giving the hypha its shape), with each compartment containing one or more nuclei; coenocytic hyphae are not compartmentalized.[42] Septa have pores that allow cytoplasm, organelles, and sometimes nuclei to pass through; an example is the dolipore septum in the fungi of the phylum Basidiomycota.[43] Coenocytic hyphae are essentially multinucleate supercells.[44]

Many species have developed specialized hyphal structures for nutrient uptake from living hosts; examples include haustoria in plant-parasitic species of most fungal phyla, and arbuscules of several mycorrhizal fungi, which penetrate into the host cells to consume nutrients.[45]

Although fungi are opisthokonts—a grouping of evolutionarily related organisms broadly characterized by a single posterior flagellum—all phyla except for the chytrids have lost their posterior flagella.[46] Fungi are unusual among the eukaryotes in having a cell wall that, in addition to glucans (e.g., β-1,3-glucan) and other typical components, also contains the biopolymer chitin.[47]

Macroscopic structures

Fungal mycelia can become visible to the naked eye, for example, on various surfaces and substrates, such as damp walls and on spoiled food, where they are commonly called molds. Mycelia grown on solid agar media in laboratory petri dishes are usually referred to as colonies. These colonies can exhibit growth shapes and colors (due to spores or pigmentation) that can be used as diagnostic features in the identification of species or groups.[48] Some individual fungal colonies can reach extraordinary dimensions and ages as in the case of a clonal colony of *Armillaria solidipes*, which extends over an area of more than 900 ha (3.5 square miles), with an estimated age of nearly 9,000 years.[49]

Armillaria solidipes

The apothecium—a specialized structure important in sexual reproduction in the ascomycetes—is a cup-shaped fruiting body that holds the hymenium, a layer of tissue containing the spore-bearing cells.[50] The fruiting bodies of the basidiomycetes (basidiocarps) and some ascomycetes can sometimes grow very large, and many are well-known as mushrooms.

Growth and physiology

The growth of fungi as hyphae on or in solid substrates or as single cells in aquatic environments is adapted for the efficient extraction of nutrients, because these growth forms have high surface area to volume ratios.[51] Hyphae are specifically adapted for growth on solid surfaces, and to invade substrates and tissues.[52] They can exert large penetrative mechanical forces; for example, the plant pathogen *Magnaporthe grisea* forms a structure called an appressorium which evolved to puncture plant tissues.[53] The pressure generated by the appressorium, directed against the plant epidermis, can exceed 8 megapascals (1200 psi).[53] The filamentous fungus *Paecilomyces lilacinus* uses a similar structure to penetrate the eggs of nematodes.[54]

Mold covering a decaying peach. The frames were taken approximately 12 hours apart over a period of six days.

The mechanical pressure exerted by the appressorium is generated from physiological processes that increase intracellular turgor by producing osmolytes such as glycerol.[55] Morphological adaptations such as these are complemented by hydrolytic enzymes secreted into the environment to digest large organic molecules—such as polysaccharides, proteins, lipids, and other organic substrates—into smaller molecules that may then be absorbed as nutrients.[56] [57] [58] The vast majority of filamentous fungi grow in a polar fashion—i.e., by extension into one direction—by elongation at the tip (apex) of the hypha.[59] Alternative forms of fungal growth include intercalary extension (i.e., by longitudinal expansion of hyphal compartments that are below the apex) as in the case of some endophytic fungi,[60] or growth by volume expansion during the development of mushroom stipes and other large organs.[61] Growth of fungi as multicellular structures consisting of somatic and reproductive cells—a feature independently evolved in animals and plants[62] —has several functions, including the development of fruiting bodies for dissemination of sexual spores (see above) and biofilms for substrate colonization and intercellular communication.[63]

Traditionally, the fungi are considered heterotrophs, organisms that rely solely on carbon fixed by other organisms for metabolism. Fungi have evolved a high degree of metabolic versatility that allows them to use a diverse range of organic substrates for growth, including simple compounds such as nitrate, ammonia, acetate, or ethanol.[64] [65] For some species it has been shown that the pigment melanin may play a role in extracting energy from ionizing radiation, such as gamma radiation; however, this form of "radiotrophic" growth has only been described for a few species, the effects on growth rates are small, and the underlying biophysical and biochemical processes are not known.[30] The authors speculate that this process might bear similarity to CO_2 fixation via visible light, but instead utilizing ionizing radiation as a source of energy.[66]

Reproduction

Fungal reproduction is complex, reflecting the differences in lifestyles and genetic makeup within this kingdom of organisms.[67] It is estimated that a third of all fungi reproduce by different modes of propagation; for example, reproduction may occur in two well-differentiated stages within the life cycle of a species, the teleomorph and the anamorph.[68] Environmental conditions trigger genetically determined developmental states that lead to the creation of specialized structures for sexual or asexual reproduction. These structures aid reproduction by efficiently dispersing spores or spore-containing propagules.

Polyporus squamosus

Asexual reproduction

Asexual reproduction via vegetative spores (conidia) or through mycelial fragmentation is common; it maintains clonal populations adapted to a specific niche, and allows more rapid dispersal than sexual reproduction.[69] The "Fungi imperfecti" (fungi lacking the perfect or sexual stage) or Deuteromycota comprise all the species which lack an observable sexual cycle.[70]

Sexual reproduction

Sexual reproduction with meiosis exists in all fungal phyla (with the exception of the Glomeromycota).<ref name=[71] It differs in many aspects from sexual reproduction in animals or plants. Differences also exist between fungal groups and can be used to discriminate species by morphological differences in sexual structures and reproductive strategies.[72] [73] Mating experiments between fungal isolates may identify species on the basis of biological species concepts.[73] The major fungal groupings have initially been delineated based on the morphology of their sexual structures and spores; for example, the spore-containing structures, asci and basidia, can be used in the identification of ascomycetes and basidiomycetes, respectively. Some species may allow mating only between individuals of opposite mating type, while others can mate and sexually reproduce with any other individual or itself. Species of the former mating system are called heterothallic, and of the latter homothallic.[74]

Most fungi have both an haploid and diploid stage in their life cycles. In sexually reproducing fungi, compatible individuals may combine by fusing their hyphae together into an interconnected network; this process, anastomosis, is required for the initiation of the sexual cycle. Ascomycetes and basidiomycetes go through a dikaryotic stage, in which the nuclei inherited from the two parents do not combine immediately after cell fusion, but remain separate in the hyphal cells (see heterokaryosis).[75]

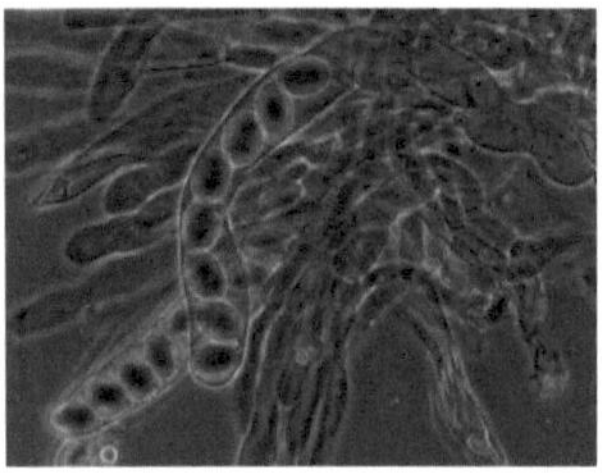

The 8-spored asci of *Morchella elata*, viewed with phase contrast microscopy

In ascomycetes, dikaryotic hyphae of the hymenium (the spore-bearing tissue layer) form a characteristic *hook* at the hyphal septum. During cell division, formation of the hook ensures proper distribution of the newly divided nuclei into the apical and basal hyphal compartments. An ascus (plural *asci*) is then formed, in which karyogamy (nuclear fusion) occurs. Asci are embedded in an ascocarp, or fruiting body. Karyogamy in the asci is followed immediately by meiosis and the production of ascospores. After dispersal, the ascospores may germinate and form a new haploid mycelium.[76]

Sexual reproduction in basidiomycetes is similar to that of the ascomycetes. Compatible haploid hyphae fuse to produce a dikaryotic mycelium. However, the dikaryotic phase is more extensive in the basidiomycetes, often also present in the vegetatively growing mycelium. A specialized anatomical structure, called a clamp connection, is formed at each hyphal septum. As with the structurally similar hook in the ascomycetes, the clamp connection in the basidiomycetes is required for controlled transfer of nuclei during cell division, to maintain the dikaryotic stage with two genetically different nuclei in each hyphal compartment.[77] A basidiocarp is formed in which club-like structures known as basidia generate haploid basidiospores after karyogamy and meiosis.[78] The most commonly known basidiocarps are mushrooms, but they may also take other forms (see Morphology section).

In glomeromycetes (formerly zygomycetes), haploid hyphae of two individuals fuse, forming a gametangium, a specialized cell structure that becomes a fertile gamete-producing cell. The gametangium develops into a zygospore, a thick-walled spore formed by the union of gametes. When the zygospore germinates, it undergoes meiosis, generating new haploid hyphae, which may then form asexual sporangiospores. These sporangiospores allow the fungus to rapidly disperse and germinate into new genetically identical haploid fungal mycelia.[79]

Spore dispersal

Both asexual and sexual spores or sporangiospores are often actively dispersed by forcible ejection from their reproductive structures. This ejection ensures exit of the spores from the reproductive structures as well as travelling through the air over long distances.

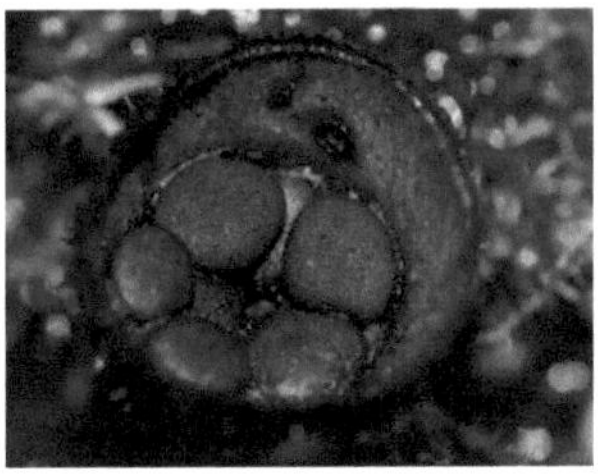

The bird's nest fungus *Cyathus stercoreus*

Specialized mechanical and physiological mechanisms, as well as spore surface structures (such as hydrophobins), enable efficient spore ejection.[80] For example, the structure of the spore-bearing cells in some ascomycete species is such that the buildup of substances affecting cell volume and fluid balance enables the explosive discharge of spores into the air.[81] The forcible discharge of single spores termed *ballistospores* involves formation of a small drop of water (Buller's drop), which upon contact with the spore leads to its projectile release with an initial acceleration of more than 10,000 g;[82] the net result is that the spore is ejected 0.01–0.02 cm, sufficient distance for it to fall through the gills or pores into the air below.[83] Other fungi, like the puffballs, rely on alternative mechanisms for spore release, such as external mechanical forces. The bird's nest fungi use the force of falling water drops to liberate the spores from cup-shaped fruiting bodies.[84] Another strategy is seen in the stinkhorns, a group of fungi with lively colors and putrid odor that attract insects to disperse their spores.[85]

Other sexual processes

Besides regular sexual reproduction with meiosis, certain fungi, such as those in the genera *Penicillium* and *Aspergillus*, may exchange genetic material via parasexual processes, initiated by anastomosis between hyphae and plasmogamy of fungal cells.[86] The frequency and relative importance of parasexual events is unclear and may be lower than other sexual processes. It is known to play a role in intraspecific hybridization[87] and is likely required for hybridization between species, which has been associated with major events in fungal evolution.[88]

Evolution

In contrast to plants and animals, the early fossil record of the fungi is meager. Factors that likely contribute to the under-representation of fungal species among fossils include the nature of fungal fruiting bodies, which are soft, fleshy, and easily degradable tissues and the microscopic dimensions of most fungal structures, which therefore are not readily evident. Fungal fossils are difficult to distinguish from those of other microbes, and are most easily identified when they resemble extant fungi.[89] Often recovered from a permineralized plant or animal host, these samples are typically studied by making thin-section preparations that can be examined with light microscopy or transmission electron microscopy.[90] Compression fossils are studied by dissolving the surrounding matrix with acid and then using light or scanning electron microscopy to examine surface details.[91]

The earliest fossils possessing features typical of fungi date to the Proterozoic eon, some 1430 [92] million years ago (Ma); these multicellular benthic organisms had filamentous structures with septa, and were capable of anastomosis.[92] More recent studies (2009) estimate the arrival of fungal organisms at about 760–1060 Ma on the basis of comparisons of the rate of evolution in closely related groups.[93] For much of the Paleozoic Era (542–251 Ma), the fungi appear to have been aquatic and consisted of organisms similar to the extant Chytrids in having flagellum-bearing spores.[94] The evolutionary adaptation from an aquatic to a terrestrial lifestyle necessitated a diversification of ecological strategies for obtaining nutrients, including parasitism, saprobism, and the development of mutualistic relationships such as mycorrhiza and lichenization.[95] Recent (2009) studies suggest that the ancestral ecological state of the Ascomycota was saprobism, and that independent lichenization events have occurred multiple times.[96]

The fungi probably colonized the land during the Cambrian (542–488.3 Ma), long before land plants.[97] Fossilized hyphae and spores recovered from the Ordovician of Wisconsin (460 Ma) resemble modern-day Glomerales, and existed at a time when the land flora likely consisted of only non-vascular bryophyte-like plants.[98] Prototaxites, which was probably a fungus or lichen, would have been the tallest organism of the late Silurian. Fungal fossils do not become common and uncontroversial until the early Devonian (416–359.2 Ma), when they are abundant in the Rhynie chert, mostly as Zygomycota and Chytridiomycota.[97] [99] [100] At about this same time, approximately 400 Ma, the Ascomycota and Basidiomycota diverged,[101] and all modern classes of fungi were present by the Late Carboniferous (Pennsylvanian, 318.1–299 Ma).[102]

Lichen-like fossils have been found in the Doushantuo Formation in southern China dating back to 635–551 Ma.[103] Lichens were a component of the early terrestrial ecosystems, and the estimated age of the oldest terrestrial lichen fossil is 400 Ma;[104] this date corresponds to the age of the oldest known sporocarp fossil, a *Paleopyrenomycites* species found in the Rhynie Chert.[105] The oldest fossil with microscopic features resembling modern-day basidiomycetes is *Palaeoancistrus*, found permineralized with a fern from the Pennsylvanian.[106] Rare in the fossil record are the homobasidiomycetes (a taxon roughly equivalent to the mushroom-producing species of the agaricomycetes). Two amber-preserved specimens provide evidence that the earliest known mushroom-forming fungi (the extinct species *Archaeomarasmius legletti*) appeared during the mid-Cretaceous, 90 Ma.[107] [108]

Some time after the Permian-Triassic extinction event (251.4 Ma), a fungal spike (originally thought to be an extraordinary abundance of fungal spores in sediments) formed, suggesting that fungi were the dominant life form at this time, representing nearly 100% of the available fossil record for this period.[109] However, the relative proportion of fungal spores relative to spores formed by algal species is difficult to assess,[110] the spike did not appear worldwide,[111] [112] and in many places it did not fall on the Permian-Triassic boundary.[113]

Taxonomy

Amoebozoa

Animalia

Choanozoa

Nucleariids

Rozellida

Microsporidia

Chytridiomycota

Neocallimastigomycota

Unikonta

Opisthokonta

Blastocladiomycota

Zoopagomycotina

Fungi[40]

Kickxellomycotina

Entomophthoromycotina

Mucoromycotina

Glomeromycota

Dikarya

Ascomycota

Basidiomycota

Although commonly included in botany curricula and textbooks, fungi are more closely related to animals than to plants and are placed with the animals in the monophyletic group of opisthokonts.[114] Analyses using molecular phylogenetics support a monophyletic origin of the Fungi.[40] The taxonomy of the Fungi is in a state of constant flux, especially due to recent research based on DNA comparisons. These current phylogenetic analyses often overturn classifications based on older and sometimes less discriminative methods based on morphological features and biological species concepts obtained from experimental matings.[115]

There is no unique generally accepted system at the higher taxonomic levels and there are frequent name changes at every level, from species upwards. Efforts among researchers are now underway to establish and encourage usage of a unified and more consistent nomenclature.[40] [116] Fungal species can also have multiple scientific names depending on their life cycle and mode (sexual or asexual) of reproduction. Web sites such as Index Fungorum and ITIS list current names of fungal species (with cross-references to older synonyms).

The 2007 classification of Kingdom Fungi is the result of a large-scale collaborative research effort involving dozens of mycologists and other scientists working on fungal taxonomy.[40] It recognizes seven phyla, two of which—the Ascomycota and the Basidiomycota—are contained within a branch representing subkingdom Dikarya. The below cladogram depicts the major fungal taxa and their relationship to opisthokont and unikont organisms. The lengths of the branches in this tree are not proportional to evolutionary distances.

Taxonomic groups

The major phyla (sometimes called divisions) of fungi have been classified mainly on the basis of characteristics of their sexual reproductive structures. Currently, seven phyla are proposed: Microsporidia, Chytridiomycota, Blastocladiomycota, Neocallimastigomycota, Glomeromycota, Ascomycota, and Basidiomycota.[40]

Phylogenetic analysis has demonstrated that the Microsporidia, unicellular parasites of animals and protists, are fairly recent and highly derived endobiotic fungi (living within the tissue of another species).[94] [117] One 2006 study concludes that the Microsporidia are a sister group to the true fungi, that is, they are each other's closest evolutionary relative.[118] Hibbett and colleagues suggest that this analysis does not clash with their classification of the Fungi, and although the Microsporidia are elevated to phylum status, it is acknowledged that further analysis is required to clarify evolutionary relationships within this group.[40]

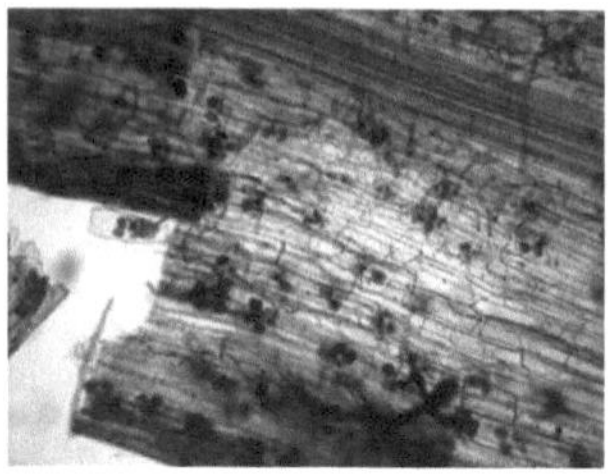

Arbuscular mycorrhiza seen under microscope. Flax root cortical cells containing paired arbuscules.

The Chytridiomycota are commonly known as chytrids. These fungi are distributed worldwide. Chytrids produce zoospores that are capable of active movement through aqueous phases with a single flagellum, leading early taxonomists to classify them as protists. Molecular phylogenies, inferred from rRNA sequences in ribosomes, suggest that the Chytrids are a basal group divergent from the other fungal phyla, consisting of four major clades with suggestive evidence for paraphyly or possibly polyphyly.[119]

The Blastocladiomycota were previously considered a taxonomic clade within the Chytridiomycota. Recent molecular data and ultrastructural characteristics, however, place the Blastocladiomycota as a sister clade to the Zygomycota, Glomeromycota, and Dikarya (Ascomycota and Basidiomycota). The blastocladiomycetes are saprotrophs, feeding on decomposing organic matter, and they are parasites of all eukaryotic groups. Unlike their close relatives, the chytrids, which mostly exhibit zygotic meiosis, the blastocladiomycetes undergo sporic meiosis.[94]

The Neocallimastigomycota were earlier placed in the phylum Chytridomycota. Members of this small phylum are anaerobic organisms, living in the digestive system of larger herbivorous mammals and possibly in other terrestrial and aquatic environments. They lack mitochondria but contain hydrogenosomes of mitochondrial origin. As the related chrytrids, neocallimastigomycetes form zoospores that are posteriorly uniflagellate or polyflagellate.[40]

Members of the Glomeromycota form arbuscular mycorrhizae, a form of symbiosis where fungal hyphae invade plant root cells and both species benefit from the resulting increased supply of nutrients. All known Glomeromycota species reproduce asexually.[71] The symbiotic association between the Glomeromycota and plants is ancient, with evidence dating to 400 million years ago.[120] Formerly part of the Zygomycota (commonly known as 'sugar' and 'pin' molds), the Glomeromycota were elevated to phylum status in 2001 and now replace the older phylum Zygomycota.[121] Fungi that were placed in the Zygomycota are now being reassigned to the Glomeromycota, or the subphyla incertae sedis Mucoromycotina, Kickxellomycotina, the Zoopagomycotina and the Entomophthoromycotina.[40] Some well-known examples of fungi formerly in the Zygomycota include black bread mold (*Rhizopus stolonifer*), and *Pilobolus* species, capable of ejecting spores several meters through the air.[122] Medically relevant genera include *Mucor*, *Rhizomucor*, and *Rhizopus*.

The Ascomycota, commonly known as sac fungi or ascomycetes, constitute the largest taxonomic group within the Eumycota.[39] These fungi form meiotic spores called ascospores, which are enclosed in a special sac-like structure called an ascus. This phylum includes morels, a few mushrooms and truffles, single-celled yeasts (e.g., of the genera *Saccharomyces*, *Kluyveromyces*, *Pichia*, and *Candida*), and many filamentous fungi living as saprotrophs, parasites, and mutualistic symbionts. Prominent and important genera of filamentous ascomycetes include *Aspergillus*, *Penicillium*, *Fusarium*, and *Claviceps*. Many ascomycete species have only been observed undergoing asexual reproduction (called anamorphic species), but analysis of molecular data has often been able to identify their closest teleomorphs in the Ascomycota.[123] Because the products of meiosis are retained within the sac-like ascus, ascomycetes have been used for elucidating principles of genetics and heredity (e.g. *Neurospora crassa*).[124]

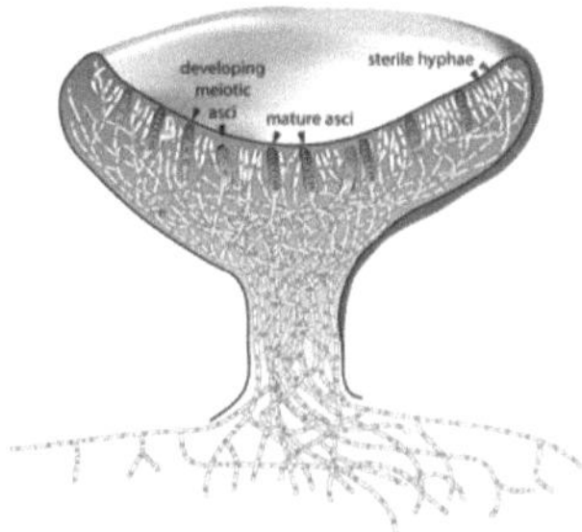

Diagram of an apothecium (the typical cup-like reproductive structure of Ascomycetes) showing sterile tissues as well as developing and mature asci.

Members of the Basidiomycota, commonly known as the club fungi or basidiomycetes, produce meiospores called basidiospores on club-like stalks called basidia. Most common mushrooms belong to this group, as well as rust and smut fungi, which are major pathogens of grains. Other important basidiomycetes include the maize pathogen *Ustilago maydis*,[125] human commensal species of the genus *Malassezia*,[126] and the opportunistic human pathogen, *Cryptococcus neoformans*.[127]

Fungus-like organisms

Because of similarities in morphology and lifestyle, the slime molds (myxomycetes) and water molds (oomycetes) were formerly classified in the kingdom Fungi. Unlike true fungi the cell walls of these organisms contain cellulose and lack chitin. Myxomycetes are unikonts like fungi, but are grouped in the Amoebozoa. Oomycetes are diploid bikonts, grouped in the Chromalveolate kingdom. Neither water molds nor slime molds are closely related to the true fungi, and, therefore, taxonomists no longer group them in the kingdom Fungi. Nonetheless, studies of the oomycetes and myxomycetes are still often included in mycology textbooks and primary research literature.[128]

The Rozellida clade, including the "chytrid" *Rozella*, is a genetically disparate group known mostly from environmental DNA sequences which is a sister group to fungi. Members of the group which have been isolated lack the chitinous cell wall which is characteristic of fungi.

The nucleariids, currently grouped in the Choanozoa, may be the next sister group to the eumycete clade, and as such could be included in an expanded fungal kingdom.[129]

Ecology

Although often inconspicuous, fungi occur in every environment on Earth and play very important roles in most ecosystems. Along with bacteria, fungi are the major decomposers in most terrestrial (and some aquatic) ecosystems, and therefore play a critical role in biogeochemical cycles[130] and in many food webs. As decomposers, they play an essential role in nutrient cycling, especially as saprotrophs and symbionts, degrading organic matter to inorganic molecules, which can then re-enter anabolic metabolic pathways in plants or other organisms.[131] [132]

Symbiosis

Many fungi have important symbiotic relationships with organisms from most if not all Kingdoms.[133] [134] [135] These interactions can be mutualistic or antagonistic in nature, or in the case of commensal fungi are of no apparent benefit or detriment to the host.[136] [137] [138]

With plants

Mycorrhizal symbiosis between plants and fungi is one of the most well-known plant–fungus associations and is of significant importance for plant growth and persistence in many ecosystems; over 90% of all plant species engage in mycorrhizal relationships with fungi and are dependent upon this relationship for survival.[139]

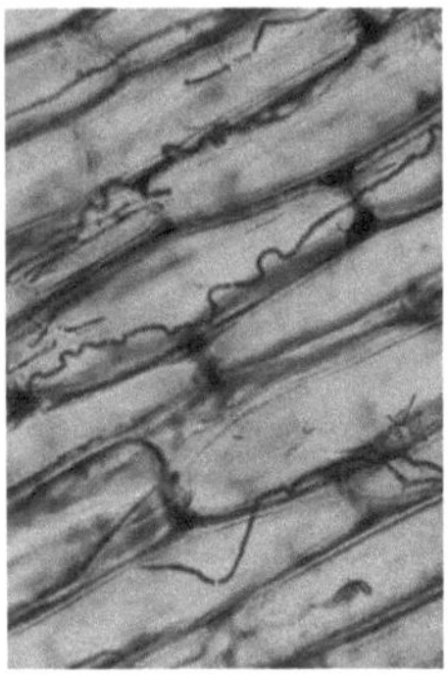

The dark filaments are hyphae of the endophytic fungus *Neotyphodium coenophialum* in the intercellular spaces of tall fescue leaf sheath tissue

The mycorrhizal symbiosis is ancient, dating to at least 400 million years ago.[120] It often increases the plant's uptake of inorganic compounds, such as nitrate and phosphate from soils having low concentrations of these key plant nutrients.[131] [140] The fungal partners may also mediate plant-to-plant transfer of carbohydrates and other nutrients. Such mycorrhizal communities are called "common mycorrhizal networks".[141] A special case of mycorrhiza is myco-heterotrophy, whereby the plant parasitizes the fungus, obtaining all of its nutrients from its fungal symbiont.[142] Some fungal species inhabit the tissues inside roots, stems, and leaves, in which case they are called endophytes.[143] Similar to mycorrhiza, endophytic colonization by fungi may benefit both symbionts; for example, endophytes of grasses impart to their host increased resistance to herbivores and other environmental stresses and receive food and shelter from the plant in return.[144]

With algae and cyanobacteria

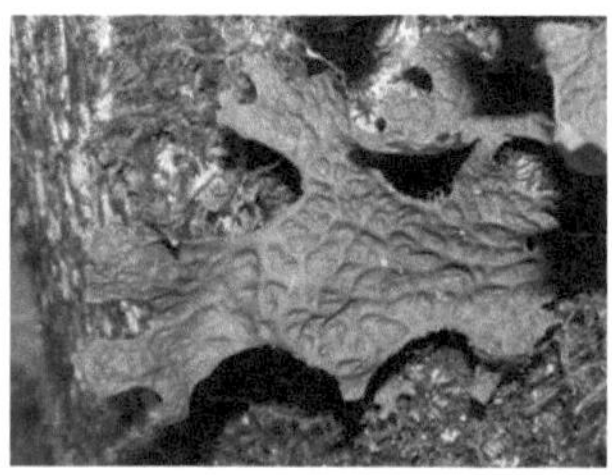

The lichen *Lobaria pulmonaria*, a symbiosis of fungal, algal, and cyanobacterial species

Lichens are formed by a symbiotic relationship between algae or cyanobacteria (referred to in lichen terminology as "photobionts") and fungi (mostly various species of ascomycetes and a few basidiomycetes), in which individual photobiont cells are embedded in a tissue formed by the fungus.[145] Lichens occur in every ecosystem on all continents, play a key role in soil formation and the initiation of biological succession,[146] and are the dominating life forms in extreme environments, including polar, alpine, and semiarid desert regions.[147] They are able to grow on inhospitable surfaces, including bare soil, rocks, tree bark, wood, shells, barnacles and leaves.[148] As in mycorrhizas, the photobiont provides sugars and other carbohydrates via photosynthesis, while the fungus provides minerals and water. The functions of both symbiotic organisms are so closely intertwined that they function almost as a single organism; in most cases the resulting organism differs greatly from the individual components. Lichenization is a common mode of nutrition; around 20% of fungi—between 17,500 and 20,000 described species—are lichenized.[149] Characteristics common to most lichens include obtaining organic carbon by photosynthesis, slow growth, small size, long life, long-lasting (seasonal) vegetative reproductive structures, mineral nutrition obtained largely from airborne sources, and greater tolerance of desiccation than most other photosynthetic organisms in the same habitat.[150]

With insects

Many insects also engage in mutualistic relationships with fungi. Several groups of ants cultivate fungi in the order Agaricales as their primary food source, while ambrosia beetles cultivate various species of fungi in the bark of trees that they infest.[151] Similarly, females of several wood wasp species (genus *Sirex*) inject their eggs together with spores of the wood-rotting fungus *Amylostereum areolatum* into the sapwood of pine trees; the growth of the fungus provides ideal nutritional conditions for the development of the wasp larvae.[152] Termites on the African savannah are also known to cultivate fungi,[133] and yeasts of the genera *Candida* and *Lachancea* inhabit the gut of a wide range of insects, including neuropterans, beetles, and cockroaches; it is not known whether these fungi benefit their hosts.[153]

As pathogens and parasites

Many fungi are parasites on plants, animals (including humans), and other fungi. Serious pathogens of many cultivated plants causing extensive damage and losses to agriculture and forestry include the rice blast fungus *Magnaporthe oryzae*,[154] tree pathogens such as *Ophiostoma ulmi* and *Ophiostoma novo-ulmi* causing Dutch elm disease,[155] and *Cryphonectria parasitica* responsible for chestnut blight,[156] and plant pathogens in the genera *Fusarium*, *Ustilago*, *Alternaria*, and *Cochliobolus*.[137] Some carnivorous fungi, like *Paecilomyces lilacinus*, are predators of nematodes, which they capture using an array of specialized structures such as constricting rings or adhesive nets.[157]

The plant pathogen *Aecidium magellanicum* causes calafate rust, seen here on a *Berberis* shrub in Chile.

Some fungi can cause serious diseases in humans, several of which may be fatal if untreated. These include aspergilloses, candidoses, coccidioidomycosis, cryptococcosis, histoplasmosis, mycetomas, and paracoccidioidomycosis. Furthermore, persons with immuno-deficiencies are particularly susceptible to disease by genera such as *Aspergillus*, *Candida*, *Cryptoccocus*,[138] [158] [159] *Histoplasma*,[160] and *Pneumocystis*.[161] Other fungi can attack eyes, nails, hair, and especially skin, the so-called dermatophytic and keratinophilic fungi, and cause local infections such as ringworm and athlete's foot.[162] Fungal spores are also a cause of allergies, and fungi from different taxonomic groups can evoke allergic reactions.[163]

Human use

The human use of fungi for food preparation or preservation and other purposes is extensive and has a long history. Mushroom farming and mushroom gathering are large industries in many countries. The study of the historical uses and sociological impact of fungi is known as ethnomycology. Because of the capacity of this group to produce an enormous range of natural products with antimicrobial or other biological activities, many species have long been used or are being developed for industrial production of antibiotics, vitamins, and anti-cancer and cholesterol-lowering drugs. More recently, methods have been developed for genetic engineering of fungi,[164] enabling metabolic engineering of fungal species. For example, genetic modification of yeast species[165] —which are easy to grow at fast rates in large fermentation vessels—has opened

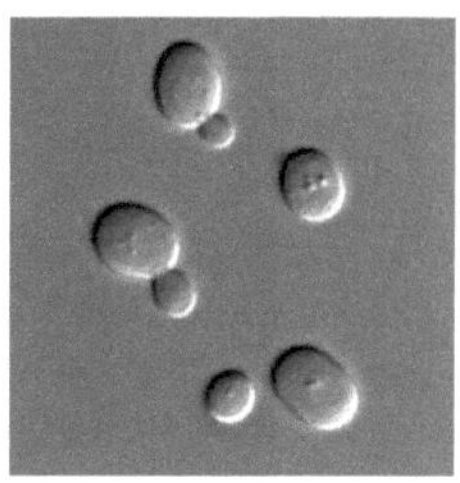

Saccharomyces cerevisiae cells shown with DIC microscopy.

up ways of pharmaceutical production that are potentially more efficient than production by the original source organisms.[166]

Drugs

Many species produce metabolites that are major sources of pharmacologically active drugs. Particularly important are the antibiotics, including the penicillins, a structurally related group of β-lactam antibiotics that are synthesized from small peptides. Although naturally occurring penicillins such as penicillin G (produced by *Penicillium chrysogenum*) have a relatively narrow spectrum of biological activity, a wide range of other penicillins can be produced by chemical modification of the natural penicillins. Modern penicillins are semisynthetic compounds, obtained initially from fermentation cultures, but then structurally altered for specific desirable properties.[167] Other antibiotics produced by fungi include: ciclosporin, commonly used as an immunosuppressant during transplant surgery; and fusidic acid, used to help control infection from methicillin-resistant *Staphylococcus aureus* bacteria.[168] Widespread use of these antibiotics for the treatment of bacterial diseases, such as tuberculosis, syphilis, leprosy, and many others began in the early 20th century and continues to play a major part in anti-bacterial chemotherapy. In nature, antibiotics of fungal or bacterial origin appear to play a dual role: at high concentrations they act as chemical defense against competition with other microorganisms in species-rich environments, such as the rhizosphere, and at low concentrations as quorum-sensing molecules for intra- or interspecies signaling.[169]

Other drugs produced by fungi include griseofulvin isolated from *Penicillium griseofulvum*, used to treat fungal infections,[170] and statins (HMG-CoA reductase inhibitors), used to inhibit cholesterol synthesis. Examples of statins found in fungi include mevastatin from *Penicillium citrinum* and lovastatin from *Aspergillus terreus* and the oyster mushroom.[171]

Cultured foods

Baker's yeast or *Saccharomyces cerevisiae*, a single-celled fungus, is used to make bread and other wheat-based products, such as pizza dough and dumplings.[172] Yeast species of the genus *Saccharomyces* are also used to produce alcoholic beverages through fermentation.[173] Shoyu koji mold (*Aspergillus oryzae*) is an essential ingredient in brewing Shoyu (soy sauce) and sake, and the preparation of miso,[174] while *Rhizopus* species are used for making tempeh.[175] Several of these fungi are domesticated species that were bred or selected according to their capacity to ferment food without producing harmful mycotoxins (see below), which are produced by very closely related *Aspergilli*.[176] Quorn, a meat substitute, is made from *Fusarium venenatum*.[177]

Medicinal use

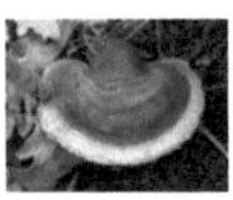

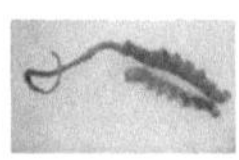

The medicinal fungi *Ganoderma lucidum* (left) and *Cordyceps sinensis* (right).

Certain mushrooms enjoy usage as therapeutics in folk medicines, such as Traditional Chinese medicine. Notable medicinal mushrooms with a well-documented history of use include *Agaricus subrufescens*,[178] [179] *Ganoderma lucidum*,[180] and *Cordyceps sinensis*.[181] Research has identified compounds produced by these and other fungi that have inhibitory biological effects against viruses[182] [183] and cancer cells.<ref name=[178] [184] Specific metabolites, such as polysaccharide-K, ergotamine, and β-lactam antibiotics, are routinely used in clinical medicine. The shiitake mushroom is a source of lentinan, a clinical drug approved for use in cancer treatments in several countries, including Japan.[185] [186] In Europe and Japan, polysaccharide-K (brand name Krestin), a chemical derived from *Trametes versicolor*, is an approved adjuvant for cancer therapy.[187]

Edible and poisonous species

Amanita phalloides accounts for the majority of fatal mushroom poisonings worldwide.

Edible mushrooms are well-known examples of fungi. Many are commercially raised, but others must be harvested from the wild. *Agaricus bisporus*, sold as button mushrooms when small or Portobello mushrooms when larger, is a commonly eaten species, used in salads, soups, and many other dishes. Many Asian fungi are commercially grown and have increased in popularity in the West. They are often available fresh in grocery stores and markets, including straw mushrooms (*Volvariella volvacea*), oyster mushrooms (*Pleurotus ostreatus*), shiitakes (*Lentinula edodes*), and enokitake (*Flammulina* spp.).[188]

There are many more mushroom species that are harvested from the wild for personal consumption or commercial sale. Milk mushrooms, morels, chanterelles, truffles, black trumpets, and *porcini* mushrooms (*Boletus edulis*) (also known as king boletes) demand a high price on the market. They are often used in gourmet dishes.[189]

Certain types of cheeses require inoculation of milk curds with fungal species that impart a unique flavor and texture to the cheese. Examples include the blue color in cheeses such as Stilton or Roquefort, which are made by inoculation with *Penicillium roqueforti*.[190] Molds used in cheese production are non-toxic and are thus safe for human consumption; however, mycotoxins (e.g., aflatoxins, roquefortine C, patulin, or others) may accumulate because of growth of other fungi during cheese ripening or storage.[191]

Stilton cheese veined with *Penicillium roqueforti*

Many mushroom species are poisonous to humans, with toxicities ranging from slight digestive problems or allergic reactions as well as hallucinations to severe organ failures and death. Genera with mushrooms containing deadly toxins include *Conocybe*, *Galerina*, *Lepiota*, and most infamously, *Amanita*.[192] The latter genus includes the destroying angel *(A. virosa)* and the death cap *(A. phalloides)*, the most common cause of deadly mushroom poisoning.[193] The false morel (*Gyromitra esculenta*) is occasionally considered a delicacy when cooked, yet can be highly toxic when eaten raw.[194] *Tricholoma equestre* was considered edible until being implicated in serious poisonings causing rhabdomyolysis.[195] Fly agaric mushrooms (*Amanita muscaria*) also cause occasional non-fatal poisonings, mostly as a result of ingestion for use as a recreational drug for its hallucinogenic properties. Historically, fly agaric was used by different peoples in Europe and Asia and its present usage for religious or shamanic purposes is reported from some ethnic groups such as the Koryak people of north-eastern Siberia.[196]

As it is difficult to accurately identify a safe mushroom without proper training and knowledge, it is often advised to assume that a wild mushroom is poisonous and not to consume it.[197] [198]

Pest control

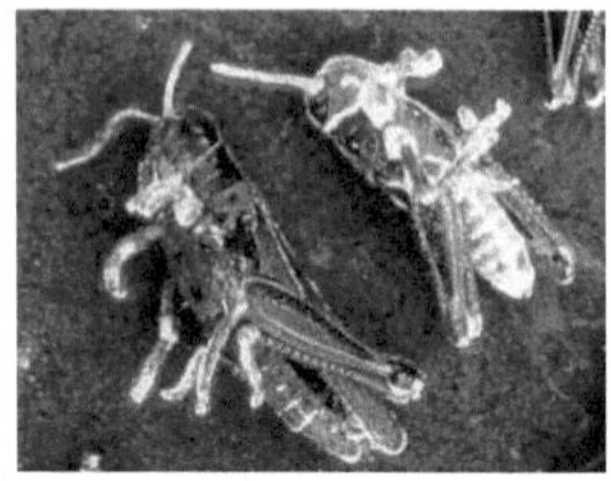
Grasshoppers killed by *Beauveria bassiana*

In agriculture, fungi may be useful if they actively compete for nutrients and space with pathogenic microorganisms such as bacteria or other fungi via the competitive exclusion principle,[199] or if they are parasites of these pathogens. For example, certain species may be used to eliminate or suppress the growth of harmful plant pathogens, such as insects, mites, weeds, nematodes and other fungi that cause diseases of important crop plants.[200] This has generated strong interest in practical applications that use these fungi in the biological control of these agricultural pests. Entomopathogenic fungi can be used as biopesticides, as they actively kill insects.[201] Examples that have been used as biological insecticides are *Beauveria bassiana*, *Metarhizium* spp, *Hirsutella* spp, *Paecilomyces* (*Isaria*) spp, and *Lecanicillium lecanii*.[202] [203] Endophytic fungi of grasses of the genus *Neotyphodium*, such as *N. coenophialum*, produce alkaloids that are toxic to a range of invertebrate and vertebrate herbivores. These alkaloids protect grass plants from herbivory, but several endophyte alkaloids can poison grazing animals, such as cattle and sheep.[204] Infecting cultivars of pasture or forage grasses with *Neotyphodium* endophytes is one approach being used in grass breeding programs; the fungal strains are selected for producing only alkaloids that increase resistance to herbivores such as insects, while being non-toxic to livestock.[205]

Bioremediation

Certain fungi, in particular "white rot" fungi, can degrade insecticides, herbicides, pentachlorophenol, creosote, coal tars, and heavy fuels and turn them into carbon dioxide, water, and basic elements.[206] Fungi have been shown to biomineralize uranium oxides, suggesting they may have application in the bioremediation of radioactively polluted sites.[207] [208] [209]

Model organisms

Several pivotal discoveries in biology were made by researchers using fungi as model organisms, that is, fungi that grow and sexually reproduce rapidly in the laboratory. For example, the one gene-one enzyme hypothesis was formulated by scientists who used the bread mold *Neurospora crassa* to test their biochemical theories.[210] Other important model fungi are *Aspergillus nidulans* and the yeasts, *Saccaromyces cerevisiae* and *Schizosaccharomyces pombe*, each of which has a long history of use to investigate issues in eukaryotic cell biology and genetics, such as cell cycle regulation, chromatin structure, and gene regulation. Other fungal models have more recently emerged that each address specific biological questions relevant to medicine, plant pathology, and industrial uses; examples include *Candida albicans*, a dimorphic, opportunistic human pathogen,[211] *Magnaporthe grisea*, a plant pathogen,[212] and *Pichia pastoris*, a yeast widely used for eukaryotic protein expression.[213]

Others

Fungi are used extensively to produce industrial chemicals like citric, gluconic, lactic, and malic acids,[214] and industrial enzymes, such as lipases used in biological detergents,[215] cellulases used in making cellulosic ethanol[216] and stonewashed jeans,[217] and amylases,[218] invertases, proteases and xylanases.[219] Several species, most notably *Psilocybin mushrooms* (colloquially known as *magic mushrooms*), are ingested for their psychedelic properties, both recreationally and religiously.

Mycotoxins

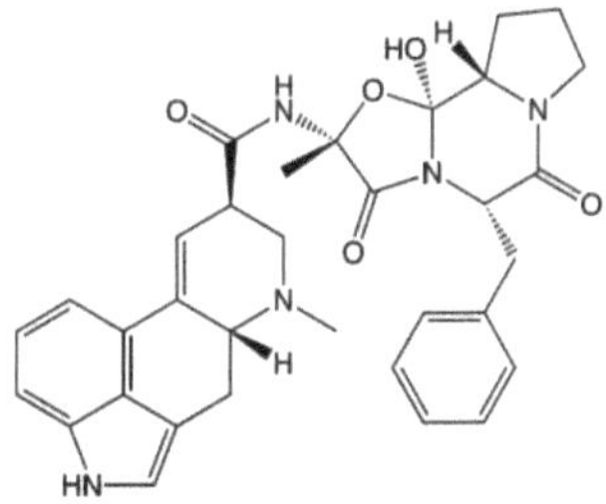
Ergotamine, a major mycotoxin produced by *Claviceps* species, which if ingested can cause gangrene, convulsions, and hallucinations

Many fungi produce biologically active compounds, several of which are toxic to animals or plants and are therefore called mycotoxins. Of particular relevance to humans are mycotoxins produced by molds causing food spoilage, and poisonous mushrooms (see above). Particularly infamous are the lethal amatoxins in some *Amanita* mushrooms, and ergot alkaloids, which have a long history of causing serious epidemics of ergotism (St Anthony's Fire) in people consuming rye or related cereals contaminated with sclerotia of the ergot fungus, *Claviceps purpurea*.[220] Other notable mycotoxins include the aflatoxins, which are insidious liver toxins and highly carcinogenic metabolites produced by certain *Aspergillus* species often growing in or on grains and nuts consumed by humans, ochratoxins, patulin, and trichothecenes (e.g., T-2 mycotoxin) and fumonisins, which have significant impact on human food supplies or animal livestock.[221]

Mycotoxins are secondary metabolites (or natural products), and research has established the existence of biochemical pathways solely for the purpose of producing mycotoxins and other natural products in fungi.[26] Mycotoxins may provide fitness benefits in terms of physiological adaptation, competition with other microbes and fungi, and protection from consumption (fungivory).[222] [223]

Mycology

Mycology is the branch of biology concerned with the systematic study of fungi, including their genetic and biochemical properties, their taxonomy, and their use to humans as a source of medicine, food, and psychotropic substances consumed for religious purposes, as well as their dangers, such as poisoning or infection. The field of phytopathology, the study of plant diseases, is closely related because many plant pathogens are fungi.[224]

Use of fungi by humans dates back to prehistory; Ötzi the Iceman, a well-preserved mummy of a 5,300 year old Neolithic man found frozen in the Austrian Alps, carried two species of polypore mushrooms that may have been used as tinder (*Fomes fomentarius*), or for medicinal purposes (*Piptoporus betulinus*).[225] Ancient peoples have used fungi as food sources–often unknowingly–for millennia, in the preparation of leavened bread and fermented juices. Some of the oldest written records contain references to the destruction of crops that were probably caused by pathogenic fungi.[226]

History

Mycology is a relatively new science that became systematic after the development of the microscope in the 16th century. Although fungal spores were first observed by Giambattista della Porta in 1588, the seminal work in the development of mycology is considered to be the publication of Pier Antonio Micheli's 1729 work *Nova plantarum genera*.[227] Micheli not only observed spores, but showed that under the proper conditions, they could be induced into growing into the same species of fungi from which they originated.[228] Extending the use of the binomial system of nomenclature introduced by Carl Linnaeus in his *Species plantarum* (1753), the Dutch Christian Hendrik Persoon (1761–1836) established the first classification of mushrooms with such skill so as to be considered a founder of modern mycology. Later, Elias Magnus Fries (1794–1878) further elaborated the classification of fungi, using spore color and various microscopic characteristics, methods still used by taxonomists today. Other notable early contributors to mycology in the 17th–19th and early 20th centuries include Miles Joseph Berkeley, August Carl Joseph Corda, Anton de Bary, the brothers Louis René and Charles Tulasne, Arthur H. R. Buller, Curtis G. Lloyd, and Pier Andrea Saccardo. The 20th century has seen a modernization of mycology that has come from advances in biochemistry, genetics, molecular biology, and biotechnology. The use of DNA sequencing technologies and phylogenetic analysis has provided new insights into fungal relationships and biodiversity, and has challenged traditional morphology-based groupings in fungal taxonomy.[229]

See also

- Conservation of fungi
- MycoBank
- Plant pathology

References

[1] Moore RT. (1980). "Taxonomic proposals for the classification of marine yeasts and other yeast-like fungi including the smuts". *Botanica Marine* **23**: 361–373.

[2] The classification system presented here is based on the 2007 phylogenetic study by Hibbett *et al.*

[3] /En-us-fungi.ogg'fʌndʒaɪ/ or /En-us-fungi-2.ogg'fʌŋɡaɪ/

[4] "Fungus" (http://www.oxforddictionaries.com/view/entry/m_en_gb0321770#m_en_gb0321770). *Oxford Dictionaries.* . Retrieved 2011-02-26.

[5] Simpson DP. (1979). *Cassell's Latin Dictionary* (5 ed.). London, UK: Cassell Ltd. p. 883. ISBN 0-304-52257-0.

[6] Ainsworth, p. 2.

[7] Alexopoulos *et al.*, p. 1.

[8] Bruns T. (2006). "Evolutionary biology: a kingdom revised". *Nature* **443** (7113): 758–761. Bibcode 2006Natur.443..758B. doi:10.1038/443758a. PMID 17051197.

[9] Baldauf; Palmer, JD (1993). "Animals and fungi are each other's closest relatives: congruent evidence from multiple proteins". *Proceedings of the National Academy of Sciences of the United States of America* **90** (24): 11558–62. Bibcode 1993PNAS...9011558B. doi:10.1073/pnas.90.24.11558. PMC 48023. PMID 8265589.

[10] Deacon, p. 4.

[11] Deacon, pp. 128–29.

[12] Alexopoulos *et al.*, pp. 28–33.

[13] Alexopoulos *et al.*, pp. 31–32.

[14] Shoji JY, Arioka M, Kitamoto K. (2006). "Possible involvement of pleiomorphic vacuolar networks in nutrient recycling in filamentous fungi". *Autophagy* **2** (3): 226–227. PMID 16874107.

[15] Deacon, p. 58.

[16] Zabriskie TM, Jackson MD. (2000). "Lysine biosynthesis and metabolism in fungi". *Natural Product Reports* **17** (1): 85–97. doi:10.1039/a801345d. PMID 10714900.

[17] Xu H, Andi B, Qian J, West AH, Cook PF. (2006). "The α-aminoadipate pathway for lysine biosynthesis in fungi". *Cellular Biochemistry and Biophysics* **46** (1): 43–64. doi:10.1385/CBB:46:1:43. PMID 16943623.

[18] Alexopoulos *et al.*, pp. 27–28.

[19] Alexopoulos *et al.*, p. 685.

[20] Desjardin DE, Oliveira AG, Stevani CV. (2008). "Fungi bioluminescence revisited". *Photochemical & Photobiological Sciences* **7** (2): 170–182. doi:10.1039/b713328f. PMID 18264584.

[21] Alexopoulos *et al.*, p. 30.

[22] Alexopoulos *et al.*, pp. 32–33.

[23] Bowman SM, Free SJ. (2006). "The structure and synthesis of the fungal cell wall". *Bioessays* **28** (8): 799–808. doi:10.1002/bies.20441. PMID 16927300.

[24] Alexopoulos *et al.*, p. 33.

[25] Mihail JD, Bruhn JN. (2005). "Foraging behaviour of *Armillaria* rhizomorph systems". *Mycological Research* **109** (11): 1195–1207. doi:10.1017/S0953756205003606. PMID 16279413.

[26] Keller NP, Turner G, Bennett JW. (2005). "Fungal secondary metabolism–from biochemistry to genomics". *Nature Reviews in Microbiology* **3** (12): 937–947. doi:10.1038/nrmicro1286. PMID 16322742.

[27] Wu S, Schalk M, Clark A, Miles RB, Coates R, Chappell J. (2007). "Redirection of cytosolic or plastidic isoprenoid precursors elevates terpene production in plants". *Nature Biotechnology* **24** (11): 1441–1447. doi:10.1038/nbt1251. PMID 17057703.

[28] Tudzynski B. (2005). "Gibberellin biosynthesis in fungi: genes, enzymes, evolution, and impact on biotechnology". *Applied Microbiology and Biotechnology* **66** (6): 597–611. doi:10.1007/s00253-004-1805-1. PMID 15578178.

[29] Vaupotic T, Veranic P, Jenoe P, Plemenitas A. (2008). "Mitochondrial mediation of environmental osmolytes discrimination during osmoadaptation in the extremely halotolerant black yeast *Hortaea werneckii*". *Fungal Genetics and Biology* **45** (6): 994–1007. doi:10.1016/j.fgb.2008.01.006. PMID 18343697.

[30] Dadachova E, Bryan RA, Huang X, Moadel T, Schweitzer AD, Aisen P, Nosanchuk JD, Casadevall A. (2007). "Ionizing radiation changes the electronic properties of melanin and enhances the growth of melanized fungi". *PLoS ONE* **2** (5): e457. Bibcode 2007PLoSO...2..457D. doi:10.1371/journal.pone.0000457. PMC 1866175. PMID 17520016.

[31] Raghukumar C, Raghukumar S. (1998). "Barotolerance of fungi isolated from deep-sea sediments of the Indian Ocean". *Aquatic Microbial Ecology* **15** (2): 153–163. doi:10.3354/ame015153.

[32] Sancho LG, de la Torre R, Horneck G, Ascaso C, de Los Rios A, Pintado A, Wierzchos J, Schuster M. (2007). "Lichens survive in space: results from the 2005 LICHENS experiment". *Astrobiology* **7** (3): 443–454. Bibcode 2007AsBio...7..443S. doi:10.1089/ast.2006.0046. PMID 17630840.

[33] Brem FM, Lips KR. (2008). "*Batrachochytrium dendrobatidis* infection patterns among Panamanian amphibian species, habitats and elevations during epizootic and enzootic stages". *Diseases of Aquatic Organisms* **81** (3): 189–202. doi:10.3354/dao01960. PMID 18998584.

[34] Le Calvez T, Burgaud G, Mahé S, Barbier G, Vandenkoornhuyse P. (2009). "Fungal diversity in deep sea hydrothermal ecosystems". *Applied and Environmental Microbiology* **75** (20): 6415–6421. doi:10.1128/AEM.00653-09. PMC 2765129. PMID 19633124.

[35] This estimation is determined by combining the species count for each phyla, based on values obtained from the 10th edition of the *Dictionary of the Fungi* (Kirk *et al.*, 2008): Ascomycota, 64163 species (p. 55); Basidiomycota, 31515 (p. 78); Blastocladiomycota, 179 (p. 94); Chytridiomycota, 706 (p. 142); Glomeromycota, 169 (p. 287); Microsporidia, >1300 (p. 427); Neocallimastigomycota, 20 (p. 463).

[36] Mueller GM, Schmit JP. (2006). "Fungal biodiversity: what do we know? What can we predict?". *Biodiversity and Conservation* **16**: 1–5. doi:10.1007/s10531-006-9117-7.

[37] Hawksworth DL. (2006). "The fungal dimension of biodiversity: magnitude, significance, and conservation". *Mycological Research* **95** (6): 641–655. doi:10.1016/S0953-7562(09)80810-1.

[38] Blackwell M. (2011). "The Fungi: 1, 2, 3 ... 5.1 million species?" (http://www.amjbot.org/cgi/reprint/98/3/333) (PDF). *American Journal of Botany* **98** (3): 426–438. doi:10.3732/ajb.1000298. PMID 21613136. .

[39] Kirk *et al.*, p. 489.

[40] Hibbett DS, Binder M, Bischoff JF, Blackwell M, Cannon PF, Eriksson OE, *et al.* (2007). "A higher level phylogenetic classification of the *Fungi*" (http://www.clarku.edu/faculty/dhibbett/AFTOL/documents/AFTOL class mss 23, 24/AFTOL CLASS MS resub.pdf) (PDF). *Mycological Research* **111** (5): 509–547. doi:10.1016/j.mycres.2007.03.004. PMID 17572334. .

[41] Harris SD. (2008). "Branching of fungal hyphae: regulation, mechanisms and comparison with other branching systems" (http://www.cybertruffle.org.uk/cyberliber/59350/0100/006/0823.htm). *Mycologia* **100** (6): 823–832. doi:10.3852/08-177. PMID 19202837. .

[42] Deacon, p. 51.

[43] Deacon, p. 57.

[44] Chang S-T, Miles PG. (2004). *Mushrooms: Cultivation, Nutritional Value, Medicinal Effect and Environmental Impact.* Boca Raton, Florida: CRC Press. ISBN 0-8493-1043-1.

[45] Parniske M. (2008). "Arbuscular mycorrhiza: the mother of plant root endosymbioses". *Nature Reviews. Microbiology* **6** (10): 763–775. doi:10.1038/nrmicro1987. PMID 18794914.

[46] Steenkamp ET, Wright J, Baldauf SL. (2006). "The protistan origins of animals and fungi" (http://mbe.oxfordjournals.org/cgi/content/full/23/1/93). *Molecular Biology and Evolution* **23** (1): 93–106. doi:10.1093/molbev/msj011. PMID 16151185. .

[47] Stevens DA, Ichinomiya M, Koshi Y, Horiuchi H. (2006). "Escape of *Candida* from caspofungin inhibition at concentrations above the MIC (paradoxical effect) accomplished by increased cell wall chitin; evidence for β-1,6-glucan synthesis inhibition by caspofungin". *Antimicrobial Agents and Chemotherapy* **50** (9): 3160–3161. doi:10.1128/AAC.00563-06. PMC 1563524. PMID 16940118.

[48] Hanson, pp. 127–41.

[49] Ferguson BA, Dreisbach TA, Parks CG, Filip GM, Schmitt CL. (2003). "Coarse-scale population structure of pathogenic *Armillaria* species in a mixed-conifer forest in the Blue Mountains of northeast Oregon". *Canadian Journal of Forest Research* **33** (4): 612–623. doi:10.1139/x03-065.

[50] Alexopoulos *et al.*, pp. 204–205.

[51] Moss ST. (1986). *The Biology of Marine Fungi.* Cambridge, UK: Cambridge University Press. p. 76. ISBN 0-521-30899-2.

[52] Peñalva MA, Arst HN. (2002). "Regulation of gene expression by ambient pH in filamentous fungi and yeasts". *Microbiology and Molecular Biology Reviews* **66** (3): 426–446. doi:10.1128/MMBR.66.3.426-446.2002. PMC 120796. PMID 12208998.

[53] Howard RJ, Ferrari MA, Roach DH, Money NP. (1991). "Penetration of hard substrates by a fungus employing enormous turgor pressures". *Proceedings of the National Academy of Sciences USA* **88** (24): 11281–11284. Bibcode 1991PNAS...8811281H. doi:10.1073/pnas.88.24.11281. PMC 53118. PMID 1837147.

[54] Money NP. (1998). "Mechanics of invasive fungal growth and the significance of turgor in plant infection". *Molecular Genetics of Host-Specific Toxins in Plant Disease: Proceedings of the 3rd Tottori International Symposium on Host-Specific Toxins, Daisen, Tottori, Japan, August 24–29, 1997.* Netherlands: Kluwer Academic Publishers. pp. 261–271. ISBN 0-7923-4981-4.

[55] Wang ZY, Jenkinson JM, Holcombe LJ, Soanes DM, Veneault-Fourrey C, Bhambra GK, Talbot NJ. (2005). "The molecular biology of appressorium turgor generation by the rice blast fungus *Magnaporthe grisea*". *Biochemical Society Transactions* **33** (Pt 2): 384–388. doi:10.1042/BST0330384. PMID 15787612.

[56] Pereira JL, Noronha EF, Miller RN, Franco OL. (2007). "Novel insights in the use of hydrolytic enzymes secreted by fungi with biotechnological potential". *Letters in Applied Microbiology* **44** (6): 573–581. doi:10.1111/j.1472-765X.2007.02151.x. PMID 17576216.

[57] Schaller M, Borelli C, Korting HC, Hube B. (2007). "Hydrolytic enzymes as virulence factors of *Candida albicans*". *Mycoses* **48** (6): 365–377. doi:10.1111/j.1439-0507.2005.01165.x. PMID 16262871.

[58] Farrar JF. (1985). "Carbohydrate metabolism in biotrophic plant pathogens". *Microbiological Sciences* **2** (10): 314–317. PMID 3939987.

[59] Fischer R, Zekert N, Takeshita N. (2008). "Polarized growth in fungi—interplay between the cytoskeleton, positional markers and membrane domains". *Molecular Microbiology* **68** (4): 813–826. doi:10.1111/j.1365-2958.2008.06193.x. PMID 18399939.

[60] Christensen MJ, Bennett RJ, Ansari HA, Koga H, Johnson RD, Bryan GT, Simpson WR, Koolaard JP, Nickless EM, Voisey CR. (2008). "*Epichloë* endophytes grow by intercalary hyphal extension in elongating grass leaves". *Fungal Genetics and Biology* **45** (2): 84–93. doi:10.1016/j.fgb.2007.07.013. PMID 17919950.

[61] Money NP. (2002). "Mushroom stem cells". *Bioessays* **24** (10): 949–952. doi:10.1002/bies.10160. PMID 12325127.

[62] Willensdorfer M. (2009). "On the evolution of differentiated multicellularity". *Evolution* **63** (2): 306–323. doi:10.1111/j.1558-5646.2008.00541.x. PMID 19154376.

[63] Daniels KJ, Srikantha T, Lockhart SR, Pujol C, Soll DR. (2006). "Opaque cells signal white cells to form biofilms in *Candida albicans*". *EMBO Journal* **25** (10): 2240–2252. doi:10.1038/sj.emboj.7601099. PMC 1462973. PMID 16628217.

[64] Marzluf GA. (1981). "Regulation of nitrogen metabolism and gene expression in fungi". *Microbiological Reviews* **45** (3): 437–461. PMC 281519. PMID 6117784.

[65] Heynes MJ. (1994). "Regulatory circuits of the *amdS* gene of *Aspergillus nidulans*". *Antonie Van Leeuwenhoek* **65** (3): 179–182. doi:10.1007/BF00871944. PMID 7847883.

[66] Dadachova E, Casadevall A. (2008). "Ionizing radiation: how fungi cope, adapt, and exploit with the help of melanin". *Current opinion in Microbiology* **11** (6): 525–531. doi:10.1016/j.mib.2008.09.013. PMC 2677413. PMID 18848901.

[67] Alexopoulos *et al.*, pp. 48–56.

[68] Kirk *et al.*, p. 633.

[69] Heitman J. (2005). "Sexual reproduction and the evolution of microbial pathogens". *Current Biology* **16** (17): R711–R725. doi:10.1016/j.cub.2006.07.064. PMID 16950098.

[70] Alcamo IE, Pommerville J. (2004). *Alcamo's Fundamentals of Microbiology*. Boston, Massachusetts: Jones and Bartlett. p. 590. ISBN 0-7637-0067-3.

[71] Redecker D, Raab P. (2006). "Phylogeny of the Glomeromycota (arbuscular mycorrhizal fungi): recent developments and new gene markers" (http://www.cybertruffle.org.uk/cyberliber/59350/0098/006/0885.htm). *Mycologia* **98** (6): 885–895. doi:10.3852/mycologia.98.6.885. PMID 17486965. .

[72] Guarro J, Gené J, Stchigel AM. (1999). "Developments in fungal taxonomy" (http://cmr.asm.org/cgi/content/full/12/3/454). *Clinical Microbiology Reviews* **12** (3): 454–500. PMC 100249. PMID 10398676. .

[73] Taylor JW, Jacobson DJ, Kroken S, Kasuga T, Geiser DM, Hibbett DS, Fisher MC. (2000). "Phylogenetic species recognition and species concepts in fungi". *Fungal Genetics and Biology* **31** (1): 21–32. doi:10.1006/fgbi.2000.1228. PMID 11118132.

[74] Metzenberg RL, Glass NL. (1990). "Mating type and mating strategies in *Neurospora*". *Bioessays* **12** (2): 53–59. doi:10.1002/bies.950120202. PMID 2140508.

[75] Jennings and Lysek, pp. 107–114.

[76] Deacon, p. 31.

[77] Alexopoulos *et al.*, pp. 492–93.

[78] Jennings and Lysek, p. 142.

[79] Deacon, pp. 21–24.

[80] Linder MB, Szilvay GR, Nakari-Setälä T, Penttilä ME. (2005). "Hydrophobins: the protein-amphiphiles of filamentous fungi". *FEMS Microbiology Reviews* **29** (5): 877–896. doi:10.1016/j.femsre.2005.01.004. PMID 16219510.

[81] Trail F. (2007). "Fungal cannons: explosive spore discharge in the Ascomycota". *FEMS Microbiology Letters* **276** (1): 12–18. doi:10.1111/j.1574-6968.2007.00900.x. PMID 17784861.

[82] Pringle A, Patek SN, Fischer M, Stolze J, Money NP. (2005). "The captured launch of a ballistospore" (http://www.cybertruffle.org.uk/cyberliber/59350/0097/004/0866.htm). *Mycologia* **97** (4): 866–871. doi:10.3852/mycologia.97.4.866. PMID 16457355. .

[83] Kirk *et al.*, p. 495.

[84] Brodie HJ. (1975). *The Bird's Nest Fungi*. Toronto, Ontario: University of Toronto Press. p. 80. ISBN 0-8020-5307-6.

[85] Alexopoulos *et al.*, p. 545.

[86] Jennings and Lysek, pp. 114–115.

[87] Furlaneto MC, Pizzirani-Kleiner AA. (1992). "Intraspecific hybridisation of *Trichoderma pseudokoningii* by anastomosis and by protoplast fusion". *FEMS Microbiology Letters* **69** (2): 191–196. doi:10.1111/j.1574-6968.1992.tb05150.x. PMID 1537549.

[88] Schardl CL, Craven KD. (2003). "Interspecific hybridization in plant-associated fungi and oomycetes: a review". *Molecular Ecology* **12** (11): 2861–2873. doi:10.1046/j.1365-294X.2003.01965.x. PMID 14629368.

[89] Donoghue MJ, Cracraft J. (2004). *Assembling the Tree of Life*. Oxford (Oxfordshire), UK: Oxford University Press. p. 187. ISBN 0-19-517234-5.

[90] Taylor and Taylor, p. 19.

[91] Taylor and Taylor, pp. 7–12.

[92] Butterfield NJ. (2005). "Probable Proterozoic fungi" (http://paleobiol.geoscienceworld.org/cgi/content/abstract/31/1/165). *Paleobiology* **31** (1): 165–182. doi:10.1666/0094-8373(2005)031<0165:PPF>2.0.CO;2. .

[93] Lucking R, Huhndorf S, Pfister D, Plata ER, Lumbsch H. (2009). "Fungi evolved right on track". *Mycologia* **101** (6): 810–822. doi:10.3852/09-016. PMID 19927746.

[94] James TY, Kauff F, Schoch CL, Matheny PB, Hofstetter V, Cox CJ, *et al.* (2006). "Reconstructing the early evolution of Fungi using a six-gene phylogeny". *Nature* **443** (7113): 818–822. Bibcode 2006Natur.443..818J. doi:10.1038/nature05110. PMID 17051209.

[95] Taylor and Taylor, pp. 84–94 and 106–107.

[96] Schoch CL, Sung GH, López-Giráldez F, Townsend JP, Miadlikowska J, Hofstetter V, *et al.* (2009). "The Ascomycota tree of life: A phylum-wide phylogeny clarifies the origin and evolution of fundamental reproductive and ecological traits". *Systematic Biology* **58** (2): 224–239. doi:10.1093/sysbio/syp020. PMID 20525580.

[97] Brundrett MC. (2002). "Coevolution of roots and mycorrhizas of land plants". *New Phytologist* **154** (2): 275–304. doi:10.1046/j.1469-8137.2002.00397.x.

[98] Redecker D, Kodner R, Graham LE. (2000). "Glomalean fungi from the Ordovician". *Science* **289** (5486): 1920–1921. Bibcode 2000Sci...289.1920R. doi:10.1126/science.289.5486.1920. PMID 10988069.

[99] Taylor TN, Taylor EL. (1996). "The distribution and interactions of some Paleozoic fungi". *Review of Palaeobotany and Palynology* **95** (1–4): 83–94. doi:10.1016/S0034-6667(96)00029-2.

[100] Dotzler N, Walker C, Krings M, Hass H, Kerp H, Taylor TN, Agerer R. (2009). "Acaulosporoid glomeromycotan spores with a germination shield from the 400-million-year-old Rhynie chert". *Mycological Progress* **8** (1): 9–18. doi:10.1007/s11557-008-0573-1.

[101] Taylor JW, Berbee ML. (2006). "Dating divergences in the Fungal Tree of Life: review and new analyses" (http://www.cybertruffle.org.uk/cyberliber/59350/0098/006/0838.htm). *Mycologia* **98** (6): 838–849. doi:10.3852/mycologia.98.6.838. PMID 17486961. .

[102] Blackwell M, Vilgalys R, James TY, Taylor JW. (2009). "Fungi. Eumycota: mushrooms, sac fungi, yeast, molds, rusts, smuts, etc." (http://tolweb.org/Fungi/2377). Tree of Life Web Project. . Retrieved 2009-04-25.

[103] Yuan X, Xiao S, Taylor TN. (2005). "Lichen-like symbiosis 600 million years ago". *Science* **308** (5724): 1017–1020. Bibcode 2005Sci...308.1017Y. doi:10.1126/science.1111347. PMID 15890881.

[104] Karatygin IV, Snigirevskaya NS, Vikulin SV. (2009). "The most ancient terrestrial lichen *Winfrenatia reticulata*: A new find and new interpretation" (http://www.springerlink.com/content/g8l8708r5gr36646/fulltext.pdf). *Paleontological Journal* **43** (1): 107–114. doi:10.1134/S0031030109010110. .

[105] Taylor TN, Hass H, Kerp H, Krings M, Hanlin RT. (2005). "Perithecial Ascomycetes from the 400 million year old Rhynie chert: an example of ancestral polymorphism" (http://www.cybertruffle.org.uk/cyberliber/59350/0097/001/0269.htm). *Mycologia* **97** (1): 269–285. doi:10.3852/mycologia.97.1.269. PMID 16389979. .

[106] Dennis RL. (1970). "A Middle Pennsylvanian basidiomycete mycelium with clamp connections" (http://www.cybertruffle.org.uk/cyberliber/59350/0062/003/0578.htm). *Mycologia* **62** (3): 578–584. doi:10.2307/3757529. JSTOR 3757529. .

[107] Hibbett DS, Grimaldi D, Donoghue MJ. (1995). "Cretaceous mushrooms in amber". *Nature* **377** (6549): 487. Bibcode 1995Natur.377..487H. doi:10.1038/377487a0.

[108] Hibbett DS, Grimaldi D, Donoghue MJ. (1997). "Fossil mushrooms from Miocene and Cretaceous ambers and the evolution of homobasidiomycetes". *American Journal of Botany* **84** (7): 981–991. doi:10.2307/2446289. JSTOR 2446289.

[109] Eshet Y, Rampino MR, Visscher H. (1995). "Fungal event and palynological record of ecological crisis and recovery across the Permian-Triassic boundary". *Geology* **23** (1): 967–970. Bibcode 1995Geo....23..967E. doi:10.1130/0091-7613(1995)023<0967:FEAPRO>2.3.CO;2.

[110] Foster CB, Stephenson MH, Marshall C, Logan GA, Greenwood PF. (2002). "A revision of Reduviasporonites Wilson 1962: description, illustration, comparison and biological affinities". *Palynology* **26** (1): 35–58. doi:10.2113/0260035.

[111] López-Gómez J, Taylor EL. (2005). "Permian-Triassic transition in Spain: a multidisciplinary approach". *Palaeogeography, Palaeoclimatology, Palaeoecology* **229** (1–2): 1–2. doi:10.1016/j.palaeo.2005.06.028.

[112] Looy CV, Twitchett RJ, Dilcher DL, Van Konijnenburg-van Cittert JHA, Visscher H. (2005). "Life in the end-Permian dead zone". *Proceedings of the National Academy of Sciences USA* **162** (4): 7879–7883. Bibcode 2001PNAS...98.7879L. doi:10.1073/pnas.131218098. PMC 35436. PMID 11427710. "See image 2".

[113] Ward PD, Botha J, Buick R, De Kock MO, Erwin DH, Garrison GH, Kirschvink JL, Smith R. (2005). "Abrupt and gradual extinction among late Permian land vertebrates in the Karoo Basin, South Africa". *Science* **307** (5710): 709–714. Bibcode 2005Sci...307..709W. doi:10.1126/science.1107068. PMID 15661973.

[114] Shalchian-Tabrizi K, Minge MA, Espelund M, Orr R, Ruden T, Jakobsen KS, Cavalier-Smith T. (2008). "Multigene phylogeny of choanozoa and the origin of animals". *PLoS ONE* **3** (5): e2098. Bibcode 2008PLoSO...3.2098S. doi:10.1371/journal.pone.0002098. PMC 2346548. PMID 18461162.

[115] See Palaeos: Fungi (http://www.palaeos.com/Fungi/default.htm) for an introduction to fungal taxonomy, including recent controversies.

[116] Celio GJ, Padamsee M, Dentinger BT, Bauer R, McLaughlin DJ. (2006). "Assembling the Fungal Tree of Life: constructing the structural and biochemical database" (http://www.cybertruffle.org.uk/cyberliber/59350/0098/006/0850.htm). *Mycologia* **98** (6): 850–859. doi:10.3852/mycologia.98.6.850. PMID 17486962. .

[117] Gill EE, Fast NM. (2006). "Assessing the microsporidia-fungi relationship: Combined phylogenetic analysis of eight genes". *Gene* **375**: 103–109. doi:10.1016/j.gene.2006.02.023. PMID 16626896.

[118] Liu YJ, Hodson MC, Hall BD. (2006). "Loss of the flagellum happened only once in the fungal lineage: phylogenetic structure of kingdom Fungi inferred from RNA polymerase II subunit genes". *BMC Evolutionary Biology* **6**: 74. doi:10.1186/1471-2148-6-74. PMC 1599754. PMID 17010206.

[119] James TY, Letcher PM, Longcore JE, Mozley-Standridge SE, Porter D, Powell MJ, Griffith GW, Vilgalys R. (2006). "A molecular phylogeny of the flagellated fungi (Chytridiomycota) and description of a new phylum (Blastocladiomycota)" (http://www.cybertruffle.org.uk/cyberliber/59350/0098/006/0860.htm). *Mycologia* **98** (6): 860–871. doi:10.3852/mycologia.98.6.860. PMID 17486963. .

[120] Remy W, Taylor TN, Hass H, Kerp H. (1994). "4-hundred million year old vesicular-arbuscular mycorrhizae". *Proceedings of the National Academy of Sciences USA* **91** (25): 11841–11843. Bibcode 1994PNAS...9111841R. doi:10.1073/pnas.91.25.11841. PMC 45331. PMID 11607500.

[121] Schüssler A, Schwarzott D, Walker C. (2001). "A new fungal phylum, the *Glomeromycota*: phylogeny and evolution". *Mycological Research* **105** (12): 1413–1421. doi:10.1017/S0953756201005196.

[122] Alexopoulos *et al.*, p. 145.

[123] For an example, see Samuels GJ. (2006). "*Trichoderma*: systematics, the sexual state, and ecology". *Phytopathology* **96** (2): 195–206. doi:10.1094/PHYTO-96-0195. PMID 18943925.

[124] Radford A, Parish JH. (1997). "The genome and genes of *Neurospora crassa*". *Fungal Genetics and Biology: FG & B* **21** (3): 258–266. doi:10.1006/fgbi.1997.0979. PMID 9290240.

[125] Valverde ME, Paredes-López O, Pataky JK, Guevara-Lara F. (1995). "Huitlacoche (*Ustilago maydis*) as a food source—biology, composition, and production". *Critical Reviews in Food Science and Nutrition* **35** (3): 191–229. doi:10.1080/10408399509527699. PMID 7632354.

[126] Zisova LG. (2009). "*Malassezia* species and seborrheic dermatitis". *Folia Medica* **51** (1): 23–33. PMID 19437895.

[127] Perfect JR. (2006). "*Cryptococcus neoformans*: the yeast that likes it hot". *FEMS Yeast Research* **6** (4): 463–468. doi:10.1111/j.1567-1364.2006.00051.x. PMID 16696642.

[128] Blackwell M, Spatafora JW. (2004). "Fungi and their allies". In Bills GF, Mueller GM, Foster MS. *Biodiversity of Fungi: Inventory and Monitoring Methods*. Amsterdam: Elsevier Academic Press. pp. 18–20. ISBN 0-12-509551-1.

[129] Shalchian-Tabrizi K, Minge MA, Espelund M, Orr R, Ruden T, Jakobsen KS, Cavalier-Smith T. (2008). "Multigene phylogeny of Choanozoa and the origin of animals". *PLoS ONE* **3** (5): e2098. Bibcode 2008PLoSO...3.2098S. doi:10.1371/journal.pone.0002098. PMC 2346548. PMID 18461162.

[130] Gadd GM. (2007). "Geomycology: biogeochemical transformations of rocks, minerals, metals and radionuclides by fungi, bioweathering and bioremediation". *Mycological Research* **111** (Pt 1): 3–49. doi:10.1016/j.mycres.2006.12.001. PMID 17307120.

[131] Lindahl BD, Ihrmark K, Boberg J, Trumbore SE, Högberg P, Stenlid J, Finlay RD. (2007). "Spatial separation of litter decomposition and mycorrhizal nitrogen uptake in a boreal forest". *New Phytologist* **173** (3): 611–620. doi:10.1111/j.1469-8137.2006.01936.x. PMID 17244056.

[132] Barea JM, Pozo MJ, Azcón R, Azcón-Aguilar C. (2005). "Microbial co-operation in the rhizosphere". *Journal of Experimental Botany* **56** (417): 1761–1778. doi:10.1093/jxb/eri197. PMID 15911555.

[133] Aanen DK. (2006). "As you reap, so shall you sow: coupling of harvesting and inoculating stabilizes the mutualism between termites and fungi". *Biology Letters* **2** (2): 209–212. doi:10.1098/rsbl.2005.0424. PMC 1618886. PMID 17148364.

[134] Nikoh N, Fukatsu T. (2000). "Interkingdom host jumping underground: phylogenetic analysis of entomoparasitic fungi of the genus *Cordyceps*". *Molecular Biology and Evolution* **17** (4): 629–638. PMID 10742053.

[135] Perotto S, Bonfante P. (1997). "Bacterial associations with mycorrhizal fungi: close and distant friends in the rhizosphere". *Trends in Microbiology* **5** (12): 496–501. doi:10.1016/S0966-842X(97)01154-2. PMID 9447662.

[136] Arnold AE, Mejía LC, Kyllo D, Rojas EI, Maynard Z, Robbins N, Herre EA. (2003). "Fungal endophytes limit pathogen damage in a tropical tree". *Proceedings of the National Academy of Sciences USA* **100** (26): 15649–15654. Bibcode 2003PNAS..10015649A. doi:10.1073/pnas.2533483100. PMC 307622. PMID 14671327.

[137] Paszkowski U. (2006). "Mutualism and parasitism: the yin and yang of plant symbioses". *Current Opinion in Plant Biology* **9** (4): 364–370. doi:10.1016/j.pbi.2006.05.008. PMID 16713732.

[138] Hube B. (2004). "From commensal to pathogen: stage- and tissue-specific gene expression of *Candida albicans*". *Current Opinion in Microbiology* **7** (4): 336–341. doi:10.1016/j.mib.2004.06.003. PMID 15288621.

[139] Bonfante P. (2003). "Plants, mycorrhizal fungi and endobacteria: a dialog among cells and genomes". *The Biological Bulletin* **204** (2): 215–220. doi:10.2307/1543562. JSTOR 1543562. PMID 12700157.

[140] van der Heijden MG, Streitwolf-Engel R, Riedl R, Siegrist S, Neudecker A, Ineichen K, Boller T, Wiemken A, Sanders IR. (2006). "The mycorrhizal contribution to plant productivity, plant nutrition and soil structure in experimental grassland". *New Phytologist* **172** (4): 739–752. doi:10.1111/j.1469-8137.2006.01862.x. PMID 17096799.

[141] Selosse MA, Richard F, He X, Simard SW. (2006). "Mycorrhizal networks: des liaisons dangereuses?". *Trends in Ecology and Evolution* **21** (11): 621–628. doi:10.1016/j.tree.2006.07.003. PMID 16843567.

[142] Merckx V, Bidartondo MI, Hynson NA. (2009). "Myco-heterotrophy: when fungi host plants". *Annals of Botany* **104** (7): 1255–1261. doi:10.1093/aob/mcp235. PMC 2778383. PMID 19767309.

[143] Schulz B, Boyle C. (2005). "The endophytic continuum". *Mycological Research* **109** (Pt 6): 661–686. doi:10.1017/S095375620500273X. PMID 16080390.

[144] Clay K, Schardl C. (2002). "Evolutionary origins and ecological consequences of endophyte symbiosis with grasses". *American Naturalist* **160** (suppl. 4): S99–S127. doi:10.1086/342161. PMID 18707456.

[145] Brodo IM, Sharnoff SD. (2001). *Lichens of North America*. New Haven, Connecticut: Yale University Press. ISBN 0-300-08249-5.

[146] Raven PH, Evert RF, Eichhorn, SE. (2005). "14—Fungi". *Biology of Plants* (7 ed.). W. H. Freeman. p. 290. ISBN 978-0-7167-1007-3.

[147] Deacon, p. 267.

[148] Purvis W. (2000). *Lichens*. Washington, D.C.: Smithsonian Institution Press in association with the Natural History Museum, London. pp. 49–75. ISBN 1-56098-879-7.

[149] Kirk *et al.*, p. 378.

[150] Deacon, pp. 267–276.

[151] Douglas AE. (1989). "Mycetocyte symbiosis in insects". *Biological Reviews of the Cambridge Philosophical Society* **64** (4): 409–434. doi:10.1111/j.1469-185X.1989.tb00682.x. PMID 2696562.

[152] Deacon, p. 277.

[153] Nguyen NH, Suh SO, Blackwell M. (2007). "Five novel *Candida* species in insect-associated yeast clades isolated from *Neuroptera* and other insects" (http://www.cybertruffle.org.uk/cyberliber/59350/0099/006/0842.htm). *Mycologia* **99** (6): 842–858. doi:10.3852/mycologia.99.6.842. PMID 18333508..

[154] Talbot NJ. (2003). "On the trail of a cereal killer: Exploring the biology of *Magnaporthe grisea*". *Annual Reviews in Microbiology* **57**: 177–202. doi:10.1146/annurev.micro.57.030502.090957. PMID 14527276.

[155] Paoletti M, Buck KW, Brasier CM. (2006). "Selective acquisition of novel mating type and vegetative incompatibility genes via interspecies gene transfer in the globally invading eukaryote *Ophiostoma novo-ulmi*". *Molecular Ecology* **15** (1): 249–262. doi:10.1111/j.1365-294X.2005.02728.x. PMID 16367844.

[156] Gryzenhout M, Wingfield BD, Wingfield MJ. (2006). "New taxonomic concepts for the important forest pathogen *Cryphonectria parasitica* and related fungi". *FEMS Microbiology Letters* **258** (2): 161–172. doi:10.1111/j.1574-6968.2006.00170.x. PMID 16640568.

[157] Yang Y, Yang E, An Z, Liu X. (2007). "Evolution of nematode-trapping cells of predatory fungi of the Orbiliaceae based on evidence from rRNA-encoding DNA and multiprotein sequences". *Proceedings of the National Academy of Sciences USA* **104** (20): 8379–8384. Bibcode 2007PNAS..104.8379Y. doi:10.1073/pnas.0702770104. PMC 1895958. PMID 17494736.

[158] Nielsen K, Heitman J. (2007). "Sex and virulence of human pathogenic fungi". *Advances in Genetics*. Advances in Genetics **57**: 143–173. doi:10.1016/S0065-2660(06)57004-X. ISBN 9780120176571. PMID 17352904.

[159] Brakhage AA. (2005). "Systemic fungal infections caused by *Aspergillus* species: epidemiology, infection process and virulence determinants". *Current Drug Targets* **6** (8): 875–886. doi:10.2174/138945005774912717. PMID 16375671.

[160] Kauffman CA. (2007). "Histoplasmosis: a clinical and laboratory update". *Clinical Microbiology Reviews* **20** (1): 115–132. doi:10.1128/CMR.00027-06. PMC 1797635. PMID 17223625.

[161] Cushion MT, Smulian AG, Slaven BE, Sesterhenn T, Arnold J, Staben C, Porollo A, Adamczak R, Meller J. (2007). "Transcriptome of *Pneumocystis carinii* during fulminate infection: carbohydrate metabolism and the concept of a compatible parasite". *PLoS ONE* **2** (5): e423. Bibcode 2007PLoSO...2..423C. doi:10.1371/journal.pone.0000423. PMC 1855432. PMID 17487271.

[162] Cook GC, Zumla AI. (2008). *Manson's Tropical Diseases: Expert Consult*. Edinburgh, Scotland: Saunders Ltd. p. 347. ISBN 1-4160-4470-1.

[163] Simon-Nobbe B, Denk U, Pöll V, Rid R, Breitenbach M. (2008). "The spectrum of fungal allergy". *International Archives of Allergy and Immunology* **145** (1): 58–86. doi:10.1159/000107578. PMID 17709917.

[164] Fincham JRS. (1989). "Transformation in fungi". *Microbiological Reviews* **53** (1): 148–170. PMC 372721. PMID 2651864.

[165] Hawkins KM, Smolke CD. (2008). "Production of benzylisoquinoline alkaloids in *Saccharomyces cerevisiae*". *Nature Chemical Biology* **4** (9): 564–573. doi:10.1038/nchembio.105. PMC 2830865. PMID 18690217.

[166] Huang B, Guo J, Yi B, Yu X, Sun L, Chen W. (2008). "Heterologous production of secondary metabolites as pharmaceuticals in *Saccharomyces cerevisiae*". *Biotechnology Letters* **30** (7): 1121–1137. doi:10.1007/s10529-008-9663-z. PMID 18512022.

[167] Brakhage AA, Spröte P, Al-Abdallah Q, Gehrke A, Plattner H, Tüncher A. (2004). "Regulation of penicillin biosynthesis in filamentous fungi". *Advances in Biochemical Engineering/Biotechnology* **88**: 45–90. doi:10.1007/b99257. PMID 15719552.

[168] Pan A, Lorenzotti S, Zoncada A. (2008). "Registered and investigational drugs for the treatment of methicillin-resistant *Staphylococcus aureus* infection". *Recent Patents on Anti-infective Drug Discovery* **3** (1): 10–33. doi:10.2174/157489108783413173. PMID 18221183.

[169] Fajardo A, Martínez JL. (2008). "Antibiotics as signals that trigger specific bacterial responses". *Current Opinion in Microbiology* **11** (2): 161–167. doi:10.1016/j.mib.2008.02.006. PMID 18373943.

[170] Loo DS. (2006). "Systemic antifungal agents: an update of established and new therapies". *Advances in Dermatology* **22**: 101–124. doi:10.1016/j.yadr.2006.07.001. PMID 17249298.

[171] Manzoni M, Rollini M. (2002). "Biosynthesis and biotechnological production of statins by filamentous fungi and application of these cholesterol-lowering drugs". *Applied Microbiology and Biotechnology* **58** (5): 555–564. doi:10.1007/s00253-002-0932-9. PMID 11956737.

[172] Kulp K. (2000). *Handbook of Cereal Science and Technology*. CRC Press. ISBN 0-8247-8294-1.

[173] Piskur J, Rozpedowska E, Polakova S, Merico A, Compagno C. (2006). "How did *Saccharomyces* evolve to become a good brewer?". *Trends in Genetics* **22** (4): 183–186. doi:10.1016/j.tig.2006.02.002. PMID 16499989.

[174] Abe K, Gomi K, Hasegawa F, Machida M. (2006). "Impact of *Aspergillus oryzae* genomics on industrial production of metabolites". *Mycopathologia* **162** (3): 143–153. doi:10.1007/s11046-006-0049-2. PMID 16944282.

[175] Hachmeister KA, Fung DY. (1993). "Tempeh: a mold-modified indigenous fermented food made from soybeans and/or cereal grains". *Critical Reviews in Microbiology* **19** (3): 137–188. doi:10.3109/10408419309113527. PMID 8267862.

[176] Jørgensen TR. (2007). "Identification and toxigenic potential of the industrially important fungi, *Aspergillus oryzae* and *Aspergillus sojae*". *Journal of Food Protection* **70** (12): 2916–2934. PMID 18095455.

[177] O'Donnell K, Cigelnik E, Casper HH. (1998). "Molecular phylogenetic, morphological, and mycotoxin data support reidentification of the Quorn mycoprotein fungus as *Fusarium venenatum*". *Fungal Genetics and Biology* **23** (1): 57–67. doi:10.1006/fgbi.1997.1018.

PMID 9501477.

[178] Hetland G, Johnson E, Lyberg T, Bernardshaw S, Tryggestad AM, Grinde B. (2008). "Effects of the medicinal mushroom *Agaricus blazei* Murill on immunity, infection and cancer". *Scandinavian Journal of Immunology* **68** (4): 363–370. doi:10.1111/j.1365-3083.2008.02156.x. PMID 18782264.

[179] Firenzuoli F, Gori L, Lombardo G. (2008). "The medicinal mushroom *Agaricus blazei* Murrill: review of literature and pharmaco-toxicological problems". *Evidence-based Complementary and Alternative Medicine: eCAM* **5** (1): 3–15. doi:10.1093/ecam/nem007. PMC 2249742. PMID 18317543.

[180] Paterson RR. (2006). "*Ganoderma*–a therapeutic fungal biofactory". *Phytochemistry* **67** (18): 1985–2001. doi:10.1002/chin.200650268. PMID 16905165.

[181] Paterson RR. (2008). "*Cordyceps*: a traditional Chinese medicine and another fungal therapeutic biofactory?". *Phytochemistry* **69** (7): 1469–1495. doi:10.1016/j.phytochem.2008.01.027. PMID 18343466.

[182] el-Mekkawy S, Meselhy MR, Nakamura N, Tezuka Y, Hattori M, Kakiuchi N, Shimotohno K, Kawahata T, Otake T. (1998). "Anti-HIV-1 and anti-HIV-1-protease substances from *Ganoderma lucidum*". *Phytochemistry* **49** (6): 1651–1657. doi:10.1016/S0031-9422(98)00254-4. PMID 9862140.

[183] El Dine RS, El Halawany AM, Ma CM, Hattori M. (2008). "Anti-HIV-1 protease activity of lanostane triterpenes from the vietnamese mushroom *Ganoderma colossum*". *Journal of Natural Products* **71** (6): 1022–1026. doi:10.1021/np8001139. PMID 18547117.

[184] Yuen JW, Gohel MD. (2005). "Anticancer effects of *Ganoderma lucidum*: a review of scientific evidence". *Nutrition and Cancer* **53** (1): 11–17. doi:10.1207/s15327914nc5301_2. PMID 16351502.

[185] Sullivan R, Smith JE, Rowan NJ. (2006). "Medicinal mushrooms and cancer therapy: translating a traditional practice into Western medicine". *Perspectives in Biology and Medicine* **49** (2): 159–170. doi:10.1353/pbm.2006.0034. PMID 16702701.

[186] Halpern GM, Miller A. (2002). *Medicinal Mushrooms: Ancient Remedies for Modern Ailments*. New York, New York: M. Evans and Co. p. 116. ISBN 0-87131-981-0.

[187] Fisher M, Yang LX. (2002). "Anticancer effects and mechanisms of polysaccharide-K (PSK): implications of cancer immunotherapy". *Anticancer Research* **22** (3): 1737–1754. PMID 12168863.

[188] Stamets P. (2000). *Growing Gourmet and Medicinal Mushrooms = [Shokuyō oyobi yakuyō kinoko no saibai]*. Berkeley, California: Ten Speed Press. pp. 233–248. ISBN 1-58008-175-4.

[189] Hall, pp. 13–26.

[190] Kinsella JE, Hwang DH. (1976). "Enzymes of *Penicillium roqueforti* involved in the biosynthesis of cheese flavor". *CRC Critical Reviews in Food Science and Nutrition* **8** (2): 191–228. doi:10.1080/10408397609527222. PMID 21770.

[191] Erdogan A, Gurses M, Sert S. (2004). "Isolation of moulds capable of producing mycotoxins from blue mouldy Tulum cheeses produced in Turkey". *International Journal of Food Microbiology* **85** (1–2): 83–85. doi:10.1016/S0168-1605(02)00485-3. PMID 12810273.

[192] Orr DB, Orr RT. (1979). *Mushrooms of Western North America*. Berkeley, California: University of California Press. p. 17. ISBN 0-520-03656-5.

[193] Vetter J. (1998). "Toxins of *Amanita phalloides*". *Toxicon* **36** (1): 13–24. doi:10.1016/S0041-0101(97)00074-3. PMID 9604278.

[194] Leathem AM, Dorran TJ. (2007). "Poisoning due to raw *Gyromitra esculenta* (false morels) west of the Rockies". *Canadian Journal of Emergency Medicine* **9** (2): 127–130. PMID 17391587.

[195] Karlson-Stiber C, Persson H. (2003). "Cytotoxic fungi—an overview". *Toxicon* **42** (4): 339–349. doi:10.1016/S0041-0101(03)00238-1. PMID 14505933.

[196] Michelot D, Melendez-Howell LM. (2003). "*Amanita muscaria*: chemistry, biology, toxicology, and ethnomycology". *Mycological Research* **107** (Pt 2): 131–146. doi:10.1017/S0953756203007305. PMID 12747324.

[197] Hall, p. 7.

[198] Ammirati JF, McKenny M, Stuntz DE. (1987). *The New Savory Wild Mushroom*. Seattle, Washington: University of Washington Press. pp. xii–xiii. ISBN 0-295-96480-4.

[199] López-Gómez J, Molina-Meyer M. (2006). "The competitive exclusion principle versus biodiversity through competitive segregation and further adaptation to spatial heterogeneities". *Theoretical Population Biology* **69** (1): 94–109. doi:10.1016/j.tpb.2005.08.004. PMID 16223517.

[200] Becker H. (1998). "Setting the Stage To Screen Biocontrol Fungi" (http://www.ars.usda.gov/is/AR/archive/jul98/fung0798.htm). United States Department of Agriculture, Agricultural Research Service. . Retrieved 2009-02-23.

[201] Keiller TS. "Whey-based fungal microfactory technology for enhanced biological pest management using fungi" (http://www.uvminnovations.com/graphics/microfactory.pdf) (PDF). UVM Innovations. . Retrieved 2009-02-23.

[202] Deshpande MV. (1999). "Mycopesticide production by fermentation: potential and challenges". *Critical Reviews in Microbiology* **25** (3): 229–243. doi:10.1080/10408419991299220. PMID 10524330.

[203] Thomas MB, Read AF. (2007). "Can fungal biopesticides control malaria?". *Nature Reviews in Microbiology* **5** (5): 377–383. doi:10.1038/nrmicro1638. PMID 17426726.

[204] Bush LP, Wilkinson HH, Schardl CL. (1997). "Bioprotective alkaloids of grass-fungal endophyte symbioses". *Plant Physiology* **114** (1): 1–7. PMC 158272. PMID 12223685.

[205] Bouton JH, Latch GCM, Hill NS, Hoveland CS, McCannc MA, Watson RH, Parish JA, Hawkins LL, Thompson FN. (2002). "Use of nonergot alkaloid-producing endophytes for alleviating tall fescue toxicosis in sheep" (http://http://jas.fass.org/content/81/5/1316.full). *Agronomy Journal* **94**: 567–574. .

[206] Christian V, Shrivastava R, Shukla D, Modi HA, Vyas BR. (2005). "Degradation of xenobiotic compounds by lignin-degrading white-rot fungi: enzymology and mechanisms involved" (http://www.ibg.uu.se/static/exjobb/99/99045.pdf) (PDF). *Indian Journal of Experimental Biology* **43** (4): 301–312. PMID 15875713. .

[207] "Fungi to fight 'toxic war zones'" (http://news.bbc.co.uk/2/hi/uk_news/scotland/tayside_and_central/7384500.stm). BBC News. May 5, 2008. .

[208] Fomina M, Charnock JM, Hillier S, Alvarez R, Gadd GM. (2007). "Fungal transformations of uranium oxides". *Environmental Microbiology* **9** (7): 1696–1710. doi:10.1111/j.1462-2920.2007.01288.x. PMID 17564604.

[209] Fomina M, Charnock JM, Hillier S, Alvarez R, Livens F, Gadd GM. (2008). "Role of fungi in the biogeochemical fate of depleted uranium". *Current Biology* **18** (9): R375–R377. doi:10.1016/j.cub.2008.03.011. PMID 18460315.

[210] Beadle GW, Tatum EL. (1941). "Genetic control of biochemical reactions in *Neurospora*" (http://www.pnas.org/content/27/11/499.short). *Proceedings of the National Academy of Sciences of the USA* **27** (11): 499–506. Bibcode 1941PNAS...27..499B. doi:10.1073/pnas.27.11.499. PMC 1078370. PMID 16588492. .

[211] Datta A, Ganesan K, Natarajan K. (1989). "Current trends in *Candida albicans* research". *Advances in Microbial Physiology*. Advances in Microbial Physiology **30**: 53–88. doi:10.1016/S0065-2911(08)60110-1. ISBN 9780120277308. PMID 2700541.

[212] Dean RA. (2005). "The genome sequence of the rice blast fungus *Magnaporthe grisea*". *Nature* **434** (7036): 980–986. Bibcode 2005Natur.434..980D. doi:10.1038/nature03449. PMID 15846337.

[213] Daly R, Hearn MT. (2005). "Expression of heterologous proteins in *Pichia pastoris*: a useful experimental tool in protein engineering and production". *Journal of Molecular Recognition: JMR* **18** (2): 119–138. doi:10.1002/jmr.687. PMID 15565717.

[214] Schlegel HG. (1993). *General Microbiology*. Cambridge, UK: Cambridge University Press. p. 360. ISBN 0-521-43980-9.

[215] Joseph B, Ramteke PW, Thomas G. (2008). "Cold active microbial lipases: some hot issues and recent developments". *Biotechnology Advances* **26** (5): 457–470. doi:10.1016/j.biotechadv.2008.05.003. PMID 18571355.

[216] Kumar R, Singh S, Singh OV. (2008). "Bioconversion of lignocellulosic biomass: biochemical and molecular perspectives". *Journal of Industrial Microbiology and Biotechnology* **35** (5): 377–391. doi:10.1007/s10295-008-0327-8. PMID 18338189.

[217] "*Trichoderma* spp., including *T. harzianum*, *T. viride*, *T. koningii*, *T. hamatum* and other spp. Deuteromycetes, Moniliales (asexual classification system)" (http://www.nysaes.cornell.edu/ent/biocontrol/pathogens/trichoderma.html). *Biological Control: A Guide to Natural Enemies in North America.* . Retrieved 2007-07-10.

[218] Olempska-Beer ZS, Merker RI, Ditto MD, DiNovi MJ. (2006). "Food-processing enzymes from recombinant microorganisms—a review". *Regulatory Toxicology and Pharmacology* **45** (2): 144–158. doi:10.1016/j.yrtph.2006.05.001. PMID 16769167.

[219] Polizeli ML, Rizzatti AC, Monti R, Terenzi HF, Jorge JA, Amorim DS. (2005). "Xylanases from fungi: properties and industrial applications". *Applied Microbiology and Biotechnology* **67** (5): 577–591. doi:10.1007/s00253-005-1904-7. PMID 15944805.

[220] Schardl CL, Panaccione DG, Tudzynski P. (2006). "Ergot alkaloids—biology and molecular biology". *The Alkaloids. Chemistry and Biology*. The Alkaloids: Chemistry and Biology **63**: 45–86. doi:10.1016/S1099-4831(06)63002-2. ISBN 9780124695634. PMID 17133714.

[221] van Egmond HP, Schothorst RC, Jonker MA. (2007). "Regulations relating to mycotoxins in food: perspectives in a global and European context". *Analytical and Bioanalytical Chemistry* **389** (1): 147–157. doi:10.1007/s00216-007-1317-9. PMID 17508207.

[222] Demain AL, Fang A. (2000). "The natural functions of secondary metabolites". *Advances in Biochemical Engineering/Biotechnology*. Advances in Biochemical Engineering/Biotechnology **69**: 1–39. doi:10.1007/3-540-44964-7_1. ISBN 978-3-540-67793-2. PMID 11036689.

[223] Rohlfs M, Albert M, Keller NP, Kempken F. (2007). "Secondary chemicals protect mould from fungivory". *Biology Letters* **3** (5): 523–525. doi:10.1098/rsbl.2007.0338. PMC 2391202. PMID 17686752.

[224] According to one 2001 estimate, some 10,000 fungal diseases are known. Struck C. (2006). "Infection strategies of plant parasitic fungi". In Cooke BM, Jones DG, Kaye B. *The Epidemiology of Plant Diseases*. Berlin, Germany: Springer. p. 117. ISBN 1-4020-4580-8.

[225] Peintner U, Pöder R, Pümpel T. (1998). "The Iceman's fungi". *Mycological Research* **102** (10): 1153–1162. doi:10.1017/S0953756298006546.

[226] Ainsworth, p. 1.

[227] Alexopoulos *et al.*, pp. 1–2.

[228] Ainsworth, p. 18.

[229] Hawksworth DL. (2006). "Pandora's Mycological Box: Molecular sequences vs. morphology in understanding fungal relationships and biodiversity". *Revista Iberoamericana de Micologia* **23** (3): 127–133. doi:10.1016/S1130-1406(06)70031-6. PMID 17196017.

Cited literature

- Ainsworth GC. (1976). *Introduction to the History of Mycology*. Cambridge, UK: Cambridge University Press. ISBN 0-521-11295-8.
- Alexopoulos CJ, Mims CW, Blackwell M. (1996). *Introductory Mycology*. John Wiley and Sons. ISBN 0-471-52229-5.
- Deacon J. (2005). *Fungal Biology*. Cambridge, Massachusetts: Blackwell Publishers. ISBN 1-4051-3066-0.
- Hall IR. (2003). *Edible and Poisonous Mushrooms of the World*. Portland, Oregon: Timber Press. ISBN 0-88192-586-1.
- Hanson JR. (2008). *The Chemistry of Fungi*. Royal Society Of Chemistry. ISBN 0-85404-136-2.
- Jennings DH, Lysek G. (1996). *Fungal Biology: Understanding the Fungal Lifestyle*. Guildford, UK: Bios Scientific Publishers Ltd. ISBN 978-1-85996-150-6.
- Kirk PM, Cannon PF, Minter DW, Stalpers JA. (2008). *Dictionary of the Fungi. 10th ed.* Wallingford, UK: CAB International. ISBN 0-85199-826-7.
- Taylor EL, Taylor TN. (1993). *The Biology and Evolution of Fossil Plants*. Englewood Cliffs, New Jersey: Prentice Hall. ISBN 0-13-651589-4.

External links

- Tree of Life web project: Fungi (http://tolweb.org/fungi)
- *Fungus* (http://www.eol.org/pages/5559) at the Encyclopedia of Life
- Mushroom Observer (mushroomobserver.org (http://mushroomobserver.org)), a collaborative fungus recording and identification project

rue:Грибы

Tick

Ticks	

Adult deer tick, *Ixodes scapularis*	
Scientific classification	
Kingdom:	Animalia
Phylum:	Arthropoda
Class:	Arachnida
Subclass:	Acari
Superorder:	Parasitiformes
Order:	**Ixodida** Leach, 1815
Superfamily:	**Ixodoidea** Leach, 1815
Families	
• Ixodidae – hard ticks • Argasidae – soft ticks • Nuttalliellidae	
Diversity	
18 genera, c. 900 species	

Tick are small arachnids in the order **Ixodida** that, along with mites, constitute the subclass Acarina. Ticks are ectoparasites (external parasites), living by hematophagy on the blood of mammals, birds, and occasionally reptiles and amphibians. Ticks are vectors of a number of diseases, including Lyme disease, Q fever (rare; more commonly transmitted by infected excreta),[1] Colorado tick fever, tularemia, tick-borne relapsing fever, babesiosis, ehrlichiosis and tick-borne meningoencephalitis, as well as bovine anaplasmosis.

Taxonomy

There are three families of ticks, one of which – Nuttalliellidae – comprises a single species, *Nuttalliella namaqua*. The remaining two families contain the hard ticks (Ixodidae) and the soft ticks (Argasidae).[2] [3]

Ixodidae are distinguished from the Argasidae by the presence of a *scutum* or hard shield. Ixodidae nymphs and adults both have a prominent *capitulum* (head) which projects forwards from the body; in the Argasidae, conversely, the capitulum is concealed beneath the body.[4] *Argasidae* contains 193 species, although the composition of the genera is less certain, and more study is needed before the genera can become stable.[2] The currently accepted genera are *Antricola*, *Argas*, *Nothaspis*, *Ornithodoros* and *Otobius*.[2]

Nuttalliella namaqua is a tick found in southern Africa from Tanzania to Namibia and South Africa,[5] which is placed in its own family, Nuttalliellidae.[2] It can be distinguished from ixodid ticks and argasid ticks by a combination of characters including the position of the stigmata, lack of setae, strongly corrugated integument, and form of the fenestrated plates.[6]

Fossilized ticks are common. Recent hypotheses based on total-evidence approach analysis place the origin of ticks in the Cretaceous (65 to 146 [7] million years ago) with most of the evolution and dispersal occurring during the Tertiary (5 to 65 [8] million years ago).[7] The oldest example is an argasid (bird) tick from Cretaceous New Jersey amber. The younger Baltic and Dominican ambers have also yielded examples, all of which can be placed in living genera.

Range and habitat

Tick species are widely distributed around the world.[8]

For an ecosystem to support ticks, it must satisfy two requirements:[9]

1. There must be a high enough population density of host species in the area.
2. There must be a high enough humidity for them to remain hydrated.

Anatomy and physiology

Ticks, like mites, have bodies which are divided into two primary sections: the anterior *capitulum* (or *gnathosoma*), which contains the head and mouthparts; and the posterior *idiosoma* which contains the legs, digestive tract, and reproductive organs.[9]

Diet and feeding behaviors

Ticks satisfy all of their nutritional requirements on a diet of blood, a practice known as hematophagy. They extract the blood by cutting a hole in the host's epidermis, into which they insert their hypostome, likely keeping the blood from clotting by excreting an anticoagulant.[10]

Ticks, such as this *Ixodes scapularis*, climb to the ends of leaves or branches and attach to hosts that brush against them.

Legs

The legs of Ixodidae and Argasidae are similar in structure. Each leg is composed of six segments: the coxa, trochanter, femur, patella, tibia, and tarsus. Each of these segments are connected by muscles which allow for flexion and extension, however the coxae have limited lateral movement. When not being used for walking, the legs remain tightly folded against the body.[11] [12]

In addition to being used for locomotion, the tarsus of leg I contains a unique sensory organ known as the Haller's organ which can detect odors and chemicals emanating from the host, as well as sensing changes in temperature and air currents.[11] [12] [13]

Life cycle and reproduction

Ixodid and argasid ticks undergo three primary stages of development: larval, nymphal, and adult.[14] Argasid ticks, however, may go through several nymphal stages, requiring a meal of blood each time.[15]

Medical issues

Tick-borne disease

Tick-borne illnesses are caused by infection with a variety of pathogens, including rickettsia and other types of bacteria, viruses, and protozoa. Because ticks can harbor more than one disease-causing agent, patients can be infected with more than one pathogen at the same time, compounding the difficulty in diagnosis and treatment. Major tick-borne diseases include Lyme disease, Rocky Mountain spotted fever, relapsing fever, tularemia, tick-borne meningoencephalitis, Colorado tick fever, Crimean-Congo hemorrhagic fever, babesiosis and cytauxzoonosis.

Eggs can be infected with pathogens inside of the ovaries, meaning that baby ticks can be infectious immediately at birth, before feeding on their first host.[15]

First aid

In general, the best way to remove adult *Ixodidae* is mechanically. If the tick's head and mouthparts are not attached to the body after removal, it may be necessary to perform a punch biopsy to remove any parts remaining inside the patient.[16]

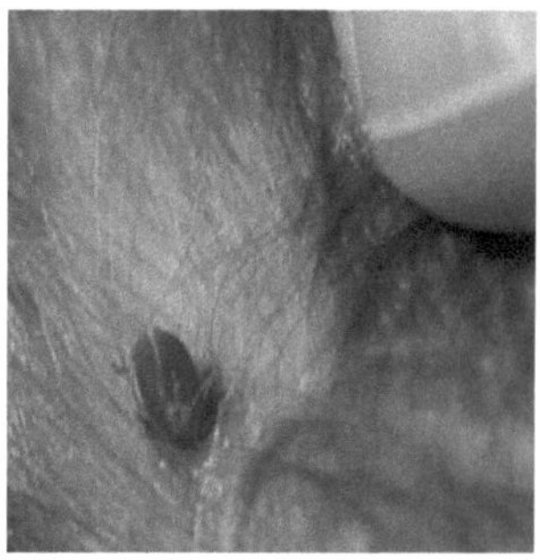

Engorged tick attached to back of toddler's head. Adult thumb shown for scale.

Population control measures

With the possible exception of widespread DDT use in the Soviet Union, attempts to limit the population or distribution of disease-causing ticks have been very unsuccessful.[17]

The parasitic Ichneumon wasp *Ixodiphagus hookeri* has long been investigated for its potential to control tick populations. It lays its eggs into ticks; the hatching wasps kill their host.

Another natural form of control for ticks is the guineafowl, a bird species which consumes mass quantities of ticks.[18] Just 2 birds can clear 2 acres (8100 m^2) in a single year.

Topical (drops/dust) flea/tick medicines may be toxic to animals and humans. Phenothrin (85.7%) in combination with Methoprene was a popular topical flea/tick therapy for felines. Phenothrin kills adult fleas and ticks. Methoprene is an insect growth regulator that interrupts the insect's life cycle by killing the eggs. However, the U.S. Environmental Protection Agency required at least one manufacturer of these products to withdraw some products and include strong cautionary statements on others, warning of adverse reactions.[19]

Case study: deer tick (*Ixodes scapularis*)

The black legged or deer tick (*Ixodes scapularis*) is dependent on the white-tailed deer for reproduction. Larval and nymph stages (immature ticks that cannot reproduce) of the deer tick feed on birds and small mammals. The adult female tick needs a large 3-day blood meal from the deer before she can reproduce and lay her 2000 or more eggs. Deer are the primary host for the adult deer tick and are key to the reproductive success of the tick.[20] See the Connecticut Agricultural Experiment Station and Connecticut Department of Public Health joint publication *Tick Management Handbook* [23] for more details of the tick's life cycle and dependence on deer.[21]

Numerous studies have shown that abundance and distribution of deer ticks are correlated with deer densities.[20] [22] [23] [24]

When the deer population was reduced by 74% at a 248-acre (100 ha) study site in Bridgeport, Connecticut, for example, the number of nymphal ticks collected at the site decreased by 92%.[20] The relationship between deer abundance, tick abundance, and human cases of Lyme disease was well documented in the Mumford Cove Community in Groton, Connecticut, from 1996 to 2004. The deer population in Mumford Cove was reduced from about 77 deer per square mile to about 10 deer per square mile (4 deer per square kilometer) after 2 years of controlled hunting. After the initial reduction the deer population was maintained at low levels. Reducing deer densities to 10 deer per square mile (4 deer per square kilometer) was adequate to reduce by more than 90% the risk of humans contracting Lyme disease in Mumford Cove.[25]

A 2006 study by Penn State's Center for Infectious Disease Dynamics indicated that reducing the deer population in small areas may lead to higher tick densities, resulting in more tick-borne infections in rodents leading to a high prevalence of tick-borne encephalitis and creating a tick hot-spot.[26]

See also

- Flea
- Mosquito
- Parasitology
- Vampire bat

Footnotes

[1] "Q fever" (http://www.cdc.gov/ncidod/dvrd/qfever). Centers for Disease Control.. Retrieved November 7, 2010.

[2] Alberto A. Guglielmone, Richard G. Robbing, Dmitry A. Apanaskevich, Trevor N. Petney, Agustín Estrada-Peña, Ivan G. Horak, Renfu Shao & Stephen C. Barker (2010). "The Argasidae, Ixodidae and Nuttalliellidae (Acari: Ixodida) of the world: a list of valid species names" (http://www.mapress.com/zootaxa/2010/f/z02528p028f.pdf) (PDF). *Zootaxa* **2528**: 1–28..

[3] Goddard, 2008: p. 80 (http://books.google.com/books?id=f-huycwyEvwC&pg=PA80)

[4] D. H. Molyneux (1993). "Vectors" (http://books.google.co.uk/books?id=jj18axV3TTAC&pg=PA6). In Francis E. G. Cox. *Modern parasitology: a textbook of parasitology* (2nd ed.). Wiley-Blackwell. pp. 53–74. ISBN 9780632025855..

[5] James E. Keirans, Carleton M. Clifford, Harry Hoogstraal & Emmett R. Easton (1976). "Discovery of *Nuttalliella namaqua* Bedford (Acarina: Ixodoidea: Nuttalliellidae) in Tanzania and redescription of the female based on scanning electron microcopy" (http://www.ingentaconnect.com/content/esa/aesa/1976/00000069/00000005/art00034). *Annals of the Entomological Society of America* **69** (5): 926–932..

[6] Mohamed A. Roshdy, Harry Hoogstraal, Abdulelah A. Banaja & Samir M. El Shoura (1983). "*Nuttalliella namaqua* (Ixodoidea: Nuttalliellidae): spiracle structure and surface morphology". *Parasitology Research* **69** (6): 817–821. doi:10.1007/BF00927431.

[7] José de la Fuente (2003). "The fossil record and the origin of ticks (Acari: Parasitiformes: Ixodida)". *Experimental and Applied Acarology* **29** (3–4): 331–334. doi:10.1023/A:1025824702816. PMID 14635818.

[8] Louis A. Magnarelli (2009). "Global importance of ticks and associated infectious disease agents". *Clinical Microbiology Newsletter* **31** (5): 33–37. doi:10.1016/j.clinmicnews.2009.02.001.

[9] Wall & Shearer, 2001: p. 55 (http://books.google.com/books?id=AMljlwB0ej0C&pg=PA55)

[10] Goddard, 2008: p. 82 (http://books.google.com/books?id=f-huycwyEvwC&pg=PA82)

[11] Sonenshine, 2005: p. 14 (http://books.google.com/books?id=dKlUARLKT9IC&pg=PA14)

[12] Nicholson *et al.*, 2009: p. 486 (http://books.google.com/books?id=6R1v9o-ual4C&pg=PA486)

[13] For Haller's organ, see also: Mehlhorn, 2008: p. 582 (http://books.google.com/books?id=Jpg1ysgVn-AC&pg=PA582).

[14] Dennis & Piesman, 2005: p. 5 (http://books.google.com/books?id=dKlUARLKT9IC&pg=PA5)
[15] Aeschlimann & Freyvogel, 1995: p. 182 (http://books.google.com/books?id=YwJF6qzhrbkC&pg=PA177)
[16] Zuber, Thomas J. & Mayeaux, E.J. (2003). "Tick Removal" (http://books.google.com/books?id=_lmEDeRGHHMC&pg=PA63). *Atlas of primary care procedurs*. Lippincott Williams & Wilkins. p. 63. ISBN 9780781739054. .
[17] Dennis & Piesman, 2005: p. 3 (http://books.google.com/books?id=dKlUARLKT9IC&pg=PA3)
[18] David Cameron Duffy, Randall Downer & Christie Brinkley (1992). "The effectiveness of Helmeted Guineafowl in the control of the deer tick, the vector of Lyme disease" (http://www.guineafowl.com/fritsfarm/guineas/ticks/tickstudy.pdf) (PDF). *Wilson Bulletin* **104** (2): 342–345. .
[19] EPA: Pesticides - Hartz Flea and Tick Drops for Cats and Kittens to be Phased Out (http://www.epa.gov/pesticides/factsheets/flea-tick-drops.htm)
[20] Stafford K.C. 2004. Tick management handbook: an integrated guide for homeowners, pest control operators, and public health officials for the prevention of tick-associated disease. The Connecticut Agricultural Experiment Station, New Haven, Connecticut, USA
[21] CAES.state.ct.us (http://www.caes.state.ct.us) p. 46, "Deer Reduction"
[22] Peter W. Rand, Charles Lubelczyk, Mary S. Holman, Eleanor H. Lacombe & Robert P. Smith, Jr. (2004). "Abundance of *Ixodes scapularis* (Acari: Ixodidae) after complete removal of deer from an isolated offshore island, endemic for Lyme disease". *Journal of Medical Entomology* **41** (4): 779–784. doi:10.1603/0022-2585-41.4.779. PMID 15311475.
[23] W. David Walter, Peter J. Perkins, Allen T. Rutberg & Howard J. Kilpatrick (2002). "Evaluation of immunocontraception in a free-ranging suburban white-tailed deer herd". *Wildlife Society Bulletin* **30** (1): 186–192. JSTOR 3784652.
[24] Mark L. Wilson, Anne M. Ducey, Thomas S. Litwin, Thomas A. Gavin & Andrew Spielman (1990). "Microgeographic distribution of immature *Ixodes dammini* ticks correlated with deer". *Medical and Veterinary Entomology* **4** (2): 151–159. doi:10.1111/j.1365-2915.1990.tb00273.x. PMID 2132979.
[25] DEP Wildlife Division: Managing Urban Deer in Connecticut 2nd edition June 2007
[26] "Deer-free areas may be haven for ticks, disease" (http://www.sciencedaily.com/releases/2006/08/060830214714.htm). Science Daily. September 4, 2006. .

References

Notes

Bibliography

- Aeschlimann, André & Thierry A. Freyvogel (1995). "Biology and distribution of ticks of medical importance" (http://books.google.com/books?id=YwJF6qzhrbkC&pg=PA177). In Jürg Meier & Julian White. *Handbook of Clinical toxicology of Animal Venoms and Poisons*. **236**. CRC Press. pp. 177–189. ISBN 9780849344893.
- Dennis, David T. & Joseph F. Piesman (2005). "Overview of tick-borne infections of humans" (http://books.google.com/books?id=dKlUARLKT9IC&pg=PA3). In Jesse L. Goodman *et al*. *Tick-borne Diseases of Humans*. ASM Press. pp. 3–11. ISBN 9781555812386.
- Goddard, Jerome (2008). "Tick-borne diseases" (http://books.google.com/books?id=f-huycwyEvwC&pg=PA80). *Infectious Diseases and Arthropods*. Springer. ISBN 9781603273992.
- Mehlhorn, Heinz & Philip M. Armstrong, ed (2001). "Ticks" (http://books.google.com/books?id=1pgvl1v_T4IC&pg=PA608). *Encyclopedic Reference of Parasitology*. Springer. pp. 608–638. ISBN 9783540668190.
- Nicholson, William L. *et al.* (2009). "Ticks (Ixodida)" (http://books.google.com/books?id=6R1v9o-ual4C&pg=PA483). In Gary Mullen & Lance Durden. *Medical and Veterinary Entomology*. Academic Press. pp. 483–532. ISBN 9780123725004.
- Sonenshine, Daniel E. (2005). "The biology of tick vectors of human disease" (http://books.google.com/books?id=dKlUARLKT9IC&pg=PA12). In Jesse L. Goodman *et al*. *Tick-borne Diseases of Humans*. ASM Press. pp. 12–36. ISBN 9781555812386.
- Wall, Richard & David Shearer (2001). "Ticks (Acari)" (http://books.google.com/books?id=AMljlwB0ej0C&pg=PA55). *Veterinary Ectoparasites: Biology, Pathology, and Control*. John Wiley & Sons. pp. 55–82. ISBN 9780632056187.

Further reading

- Bruin, Jan & L. P. S. van der Greest, ed (2009). *Diseases of Mites and Ticks* (http://books.google.com/books?id=2gz_JSTOzZ8C). Springer. ISBN 9781402096945.
- Cunha, Burke A. (2000). *Tickborne Infectious Diseases: Diagnosis and Management* (http://books.google.com/books?id=UOKub5HbpoYC). CRC Press. ISBN 9780824703103.
- Service, M. W. (2004). "Soft ticks (Argasidae)" (http://books.google.com/books?id=yi5gRCErrrQC&pg=PA213). *Medical Entomology for Students*. Cambridge University Press. pp. 213–221. ISBN 9780521547758.
- Service, M. W. (2004). "Hard ticks (Ixodidae)" (http://books.google.com/books?id=yi5gRCErrrQC&pg=PA222). *Medical Entomology for Students*. Cambridge University Press. pp. 222–236. ISBN 9780521547758.
- Sonenshine, Daniel E. (1993). *Biology of Ticks* (http://books.google.com/books?id=6LXwAAAAMAAJ). **Vols. 1 & 2**. Oxford University Press. ISBN 9780195084313.

Article Sources and Contributors

Foreign grain beetle *Source*: http://en.wikipedia.org/w/index.php?title=Foreign_grain_beetle *Contributors*: Chris the speller, Chuhymen356, Dawynn, Finngall, Jangirl221, Kansan, KrakatoaKatie, LOL, Leebo, LilHelpa, Linjacqu, Marwood, Mm40, Mr Insect, Rjwilmsi, Stemonitis, Stepshep, TenPoundHammer, Tinglejo, 3 anonymous edits

Beetle *Source*: http://en.wikipedia.org/w/index.php?title=Beetle *Contributors*: (jarbarf), 157apt, 162.129.26.xxx, 2112 rush, 417.417, 7, 80.255, 99of9, Abce2, Abigail-II, Abyssal, Acalamari, Ace ETP, AckleyJ, Acroterion, Addshore, Adrian.benko, Ahoerstemeier, Aitias, Akylea, Alansohn, Alexnnnv1, Alison, Amtin, Andareed, Andre Engels, Andycjp, Animum, Anoopkn, Antandrus, Anypodetos, Arachn0, Arandia, Arpingstone, Astral, Autumn Attic, AxelStrauss, BD2412, Bagatelle, Baldhur, BananaFiend, Bart133, Beetledude2, Bendzh, Bensaccount, Bettia, Bhadani, Bidgee, Binary TSO, Bkell, Bkkbrad, Blcfilm, Bob the Wikipedian, Bobo192, Bodnotbod, Boing! said Zebedee, Bonadea, Bongwarrior, Borgx, Boris Barowski, Brandon5485, BraneJ, Brian8710, BrianAdler, Brighterorange, Brograve, Brookie, Bruce Marlin, Bugboy52.40, Bugpoo, CIreland, Calliopejen1, Can't sleep, clown will eat me, Capnez, Capricorn42, CardinalDan, Celithemis, Chris the speller, Chuck Entz, Cinik, Clayoquot, Cliff smith, Coasterlover1994, Comppsi, Connie Chiwa, Connorwithteeth2, Conscious, Conversion script, Cookieface1234, Cotinis, Courcelles, CtrlAltDek, Cwmhiraeth, Cyanolinguophile, DJ Clayworth, DVD R W, Da monster under your bed, DarkFalls, Darth Vader is your Father, David Eppstein, DavidLevinson, Dcooper, DeadEyeArrow, Deadlymachine23, Deflective, Densock, Deor, DerBorg, Dinardop356, DirkvdM, Discospinster, Dolovis, DopefishJustin, Download, Dr. Günter Bechly, Dthomsen8, DuKot, Durova, Dyanega, Dylan620, Dysepsion, ERcheck, EWhitted, Ebeisher, Eclecticology, Ecphora, Edderso, Electric goat, Eluchil404, Emperorbma, Enchanter, EncycloPetey, Enviroboy, Epbr123, Ernyerny, Esanchez7587, Esn, Ettrig, Etxrge, Eumolpo, Evil Monkey, FF2010, Fartessimo, Fastilysock, FelineAvenger, Fir0002, Florentino floro, Fluffycoffe, Forgiveme997, Friesse, Frodjet Wraith, Frozenevolution, Fudoreaper, Gaius Cornelius, Gdr, GeorgieParr, Gephart, Gigemag76, Gilliam, Giraffedata, Glane23, Glenfarclas, Gogo Dodo, Goliathus, Good day111, Gordonjcp, Graham87, Grahbudd, Grayshi, Gscshoyru, Gtorell, Gutsul, Guy M, Gzuckier, Happy5214, Heds, Hellbus, Hephaestos, Hottentot, Hugo.arg, Hydro, Hzglenn, IRP, Iketsi, Immunize, Infrogmation, Insectpollinators, IronChris, Isnow, Ixfd64, J.delanoy, JForget, JWSchmidt, JamesAM, Jason M, Jebba, Jebus989, Jecowa, Jeekc, Jeffrey O. Gustafson, Jhml, Jhoveson, Jhsounds, Jiffijosh, JimDunning, Jimbohendrix, Jimp, Jmundo, Jnagyjr, Jncraton, JoJan, John of Reading, John-Nash, John254, JohnCD, JohnOwens, Johnson1333, Jovianeye, Jpdnnet, Jpgordon, Jsavard, Jtflood1976, Jung hyuk, Jwissick, K.C. Tang, Kaarel, Kamranki, KathrynLybarger, Keith Edkins, Ken Gallager, Kerotan, Kettleys, Kevin Ryde, KhaTzek, Khoikhoi, King of Hearts, Kinkyturnip, Kjoonlee, Kmehlin, Kmoksy, KnowledgeOfSelf, Kodd, Kpts44, Krissiie, Kubigula, Kugamazog, Kurykh, Kwamikagami, Ladybug63, Landon1980, Laurifer, Lcarscad, Leevanjackson, Lexor, Lights, Limbo socrates, LindsayH, Lollies 1212, Lotje, Luk, Luna Santin, MER-C, MK8, MPF, MTorzsa1456, Macarooni, Magioladitis, Magnus Manske, Malleus Fatuorum, Maltesedog, Mani1, Manuel Anastácio, Marc Venot, MarkSutton, Marknen, Mav, Mbz1, McSly, Mclay1, Mcwbr, Medeis, Melaen, Menchi, Mendaliv, Mentifisto, Mgiganteus1, Minesweeper, Minna Sora no Shita, Miss Madeline, Missylisa153, Mnemeson, Mocahantas, Modulok, Monkeymanman, Moshe Constantine Hassan Al-Silverburg, Mr.pieface, Mrbines, Mrintel, Mu2, Muhammad Mahdi Karim, Muke, Mulad, Mweites, My Core Competency is Competency, Mysid, N5iln, Naddy, Nanodance, Netoholic, Nihiltres, Nikai, Nilsaxel, Nivix, Notafly, Obentomusubi, Oda Mari, Opelio, Oroso, PaddyM, PaperTruths, Pavel Vozenilek, Peter Clausen, Peter coxhead, PeterKarels, Petrb, Phearson, Philip Trueman, Phy1729, Piano non troppo, PiccoloNamek, PierreAbbat, Pigsonthewing, Pinethicket, Pinky sl, Poindexter Propellerhead, Pollinator, Prodego, ProudWestVirginiaNative, Proyster, Pseudomonas, PuzzletChung, Quangbao, Qwertt900, RZINZ, Racheldk1, RaseaC, Rawlife, Redyka94, Reedy, Relicsmaster, Reneeholle, RexNL, Rhobite, Rhopkins8, Rich Farmbrough, RichG, Richard001, Rickmansbrick, Rjwilmsi, Robert Skyhawk, Ronhjones, Roregan, Rotational, Rumping, Runewiki777, SP-KP, Salamander455, Sam Hocevar, Samsara, Scapler, Scoops, Seb az86556, Sengkang, Sesshomaru, Sgassett, Sharkman, Shenme, Shyamal, Slon02, Smallweed, Snaketastic, Snigbrook, Snowolf, Sobriquet, Someone else, Sonicrazy, Spalding, Spellcast, Squash, Stan Shebs, Stemonitis, Stephen C. Carlson, Studerby, Stwalkerster, Svjetlana3, Sylverfysh, Sylwia Ufnalska, TUF-KAT, Tameeria, Tannin, Tarquin, Tavilis, Teflon Don, Template namespace initialisation script, That Guy, From That Show!, The Thing That Should Not Be, The inquisitor lord, Tiddly Tom, Tide rolls, TigerShark, Tiptoety, Tnxman307, Toytoy, Traxs7, Treepugger, Treisijs, Trfasulo, Twc55, Twerbrou, Tyrenius, UpstateNYer, Urgabne, User2004, Ute in DC, UtherSRG, Vague Rant, Vicki Rosenzweig, Vratislav, WHARRGARBL, WPIsFlawed, Wakedream, Wavelength, Weird Bird, WikiCopter, WikipedianProlific, Wikipelli, William Avery, Williamb, Wknight94, Woohookitty, WormRunner, Wykily, Xezbeth, Xvazquez, XxNotoriousxX, Youshouldask, ZICO, Zeimusu, 721 anonymous edits

Oryzaephilus surinamensis *Source*: http://en.wikipedia.org/w/index.php?title=Oryzaephilus_surinamensis *Contributors*: Andybrookestar, Art LaPella, Badgernet, Computer room guy, Dawynn, Heds, Kugamazog, MECU, Sarefo, Stemonitis, Trfasulo, 5 anonymous edits

Flea *Source*: http://en.wikipedia.org/w/index.php?title=Flea *Contributors*: 1dragon, 28bytes, 2help, 2over0, 3bigbass, 4twenty42o, 902unfortunate902, A little insignificant, Abanima, Abigail-II, Abrahamschulte, Adam78, AdjustShift, Aesop32809, Aflm, Airplaneman, Alansohn, Allstarecho, Amcfreely, Ameliabedelia546, Anaxial, Andrarias, AndrewHowse, Andrewlp1991, Animum, Aniten21, Anjana 04, AnmaFinotera, Anna555, Anomie, Antandrus, Anteaterpest, Apokryltaros, Apostrophe, Arcadian, Ardonik, Arkuat, Arpingstone, AshLin, Ashcraft, Astrozyll, Avjoska, BG65, Ballroom16, Bart133, Baseballrocks538, BeautifulFlying, Beetstra, Belchman, Ben Hateva, Bender235, Bihco, Binarypower, Blahedo, Bob98133, Bobo192, Bojars, Bookermorgan, Bootlog, Brain40, Brambleshire, BruceGannon, Bruinfan12, C-3PO, CWii, Cacycle, CalamusFortis, Caleb Murdock, Calvin 1998, Cam1309, Cambrasa, CanadianLinuxUser, CapeVerdeWave, CaritasUbi, Cavell, Cgaffney, Chas zzz brown, Chew-Zilla, Chris 73, Christophenstein, Chun-hian, Clam0p, Clawed, Cliffb, Cobaltcigs, Collegentravel, Cometstyles, Corinthian, Correogsk, CrazyLegsKC, Crazytales, Da monster under your bed, Damir A, Danmohammad, Darth Panda, DarthShrine, Debresser, Delirium, DerBorg, Diligent Terrier, Djsolie, DocWatson42, Docelder, Doctorfluffy, Doggiedog, Doseiai2, Dpup77, Dr Zak, DragonZero, Drakejj, Dukeofomnium, Dyanega, Dysmorodrepanis, Eb.eric, Eclecticology, Eeekster, El C, Elf, Eli lilly, Epbr123, Eric Forste, Eric-Wester, ErikNY, Esanchez7587, Esculapio, Euph, Excirial, Fabartus, Falcon8765, Fanra, Filipo67, Fleebag, Flowerpotman, Flukeboy, Fongs, Frank Warmerdam, FrickaFricka, Furrykef, Gaff, Gail, Gauss, Gothprincess113, Grant M, Gront, Ground Zero, Gucciii, Gurch, Hadal, Headbomb, Hede2000, Hephaestos, Herbmandan, Here.it.comes.again, Hydro, Icep, Irishguy, IronChris, Isison, It Is Me Here, Ivirivi00, J.delanoy, JNW, JaGa, JamesMLane, Jauhienij, Jawju, Jeffrey Mall, Jidanni, Jimfbleak, Joelmills, Johnrs1, Jonkerz, Jordanvendramin, Joshua Scott, Jpgordon, Jraephelan, JulliaWhite, Kar.ma, Kc0yef, Kent Wang, Khoikhoi, KikuichimonjiEX, Kmoksy, Koavf, Koven.rm, Kubigula, Kugamazog, Kuru, L Kensington, LOL, Lachatdelarue, Lascorz, Lexco123, Lightmouse, LittleDan, Local-Search-Results, Lockesdonkey, Lshiner, Luna Santin, Lyndsayruell, M karzarj, MER-C, Macaddct1984, Maelor, Magore, Malo, Mandrake76, Mangajunglist, Marc Venot, Marcus Qwertyus, Marek69, Mason1024, MattieTK, Maximus Rex, McDogm, McSly, Mechanical digger, Mercapatan101, Merotoker1, Mieciu K, Mikeaponte1, Mirrordor, Mr. Billion, MrDarwin, MrHighway, Mud, Muriel Gottrop, Mushroom, Mwanner, My Core Competency is Competency, NHRHS2010, NPIC, Nanodance, Nerwen, NewEnglandYankee, Nihiltres, Noca2plus, Nono64, Northumbrian, Numen, Nurnord, Nzfooty, Oasisbob, Observer31, Ohnoitsjamie, Olivercatterson, OneAmongBillions, Oneiros, Outesticide, OwenX, PDH, PTJoshua, Paolo.dL, PatrickFisher, Pedro, Pekinensis, Peruvianllama, Peter Grey, Pgan002, Philip Trueman, Piano non troppo, PierreAbbat, Piledhigheranddeeper, Pollinator, Potatoswatter, Power piglet, Pscriba, Pulsar.co.nr, Pyrope, RHCP1423!, RedWolf, Redorganica, Reed.lacy, Renamed user 33, Res2216firestar, RexNL, Rhoganjosh, Richard001, RickK, RobertG, Ronhjones, Rotational, Rpyle731, Ruhrfisch, Runwild2006, Samb008, Samoya, Sarefo, Sbharris, SchmuckyTheCat, Sdcrym, Seb az86556, Seduisant, Semidriver214, Severoon, SharperUSHER 2, Shirik, Shoeofdeath, Shyamal, SimonD, Sinevandor44, SiobhanHansa, Sjorford, Snowmanradio, Snowmonster, Sonimom, Sparr00w, SpecMode, Specious, Stambouliote, Stefan, Steff2, Stemonitis, Stephenb, SteveBaker, Stevenwmccrary58, Styrofoam1994, Sue0189, Syra987, SystemBuilder, Szquirrel, Takarada, Tarquin, Taxman, Template namespace initialisation script, The Sanctuary Sparrow, TheOtherJesse, Theo480, Thisisbossi, Thndr333, Tide rolls, Tim Q. Wells, Tim1357, TimonyCrickets, Tommy2010, Topbanana, Treastermj, Trfasulo, Trysha, Turian, TwoOneTwo, U+003F, Ubardak, UtherSRG, Victorcoutin, Vipinhari, WODUP, WadeSimMiser, Wavelength, Wayne Roberson, Austin, Texas, West.andrew.g, Wiki alf, Williamb, Winchelsea, Windizy, Winston365, Woohookitty, Wrelwser43, Yamamoto Ichiro, Yath, Yelyos, Yerpo, Yovoyovo, Zelbinian, ZrikiSvargla, 653 anonymous edits

Drosophila melanogaster *Source*: http://en.wikipedia.org/w/index.php?title=Drosophila_melanogaster *Contributors*: 123dylan456, 1wernetj, Aciel, AdamRetchless, Addshore, Adrian J. Hunter, Agathman, Aka, Alan Liefting, Alboyle, Albval, Alex1011, Alleis, Alvis, AndreasJS, AndyKali, Anomalocaris, AnteaterZot, AstarothCY, AxelBoldt, Azhyd, B.d.mills, Baskaufs, BeefRendang, Biederman, Blake Burba, Blue520, Bobianite, Bobo192, Bobrayner, BuddhaBubba, Bugboy52.40, Caltas, CambridgeBayWeather, Casliber, Cbs30954, Celefin, Charles Matthews, Chcknwnm, Chinasaur, Chris Roy, Clock39, Cohesion, Conops, Coolhandscot, Crunkologist, Crustaceanguy, Dadude3320, DanBlick, Darth Panda, David D., Debivort, DennyColt, Digitalgadget, Diliff, Dina, DocWatson42, Doodle77, Doprendek, Duncharris, EdJohnston, Egan86, Egfr, Ekyygork, El Beadur, Emefectivo, Emw, Ettrig, FJPB, Favonian, Fireaxe888, Fireworks, Flopster2, Floydspinky71, Flyguy649, Flyskippy1, Foggg, Frankenpuppy, Gabbe, Gaius Cornelius, Gaurav, Gavin.perch, Gbleem, Giftlite, Gilliam, Glenn, Gruzd, Gurch, Gwguffey, Hadavid52, HalfShadow, Heds, HiLo48, Hokiedokie, Hottentot, Hybridstyle, Ian1, IslandGyrl, Isnow, Ivo, J.delanoy, JLD, JWSchmidt, Jamesofur, JeremyR, Jilongliu, Jim Douglas, Jim.henderson, Jmrowland, John Hill, JohnJohn, JorgeGG, Jose Ramos, Jredmond, Jsonitsac, Judgeking, Junkyardprince, Jusdafax, KarlM, Keegscee, Khaosworks, Kimon, KimvdLinde, Kmcallenberg, Knowledge Seeker, KnowledgeOfSelf, Kozuch, Kpjas, Kristenq, Kungfuadam, Kupirijo, L Kensington, Lawrence Cohen, Lexor, Ligulem-s, Lostchicken, LouScheffer, Loueylewis, M1ss1ontomars2k4, MPF, Magister Mathematicae, Malkinann, Maniago, MarcoTolo, Marj Tiefert, Mark-mitchell-aldershot, MarkBuckles, Marshman, MastCell, Maxim, Mgiganteus1, MiPe, Mike Lin, MikeLynch, Mild Bill Hiccup, Mimihitam, Minimac, MisterSheik, Mnyakaba-GMU, Mrsgensho, Murgh, Muriel Gottrop, N3VeS, NGC6254, Nabla, Naj-GMU, Newton2, Nick in syd, Notafly, OhanaUnited, OllieFury, Omicronpersei8, Ost316, P1h3r1e3d13, PDH, Parasite, Perfecto, Peterfarez, Philip Trueman, Phoenix Flower, Piltro, Piolinfax, Plindenbaum, Plm209, Pne, Pollinator, Possum, Proquence, Pseudomonas, Qartis, Quiddity, RDBrown, Raul654, Rch14, RecercaenAccio, RexNL, Rich Farmbrough, Richard001, Rickhanulewicz, Rickproser, Rjwilmsi, Rothrube, Rsabbatini, Rui278, Salvor, Saperaud, Sarefo, SchuminWeb, Setanta747 (locked), ShadowPuppet, Sjschen, SkyBlue eagle, Smartse, Snilsen, Socialservice, Someone42, Sonimom, Speciate, Speh, Spencer, Spindled, Stan Shebs, Stemonitis, Steveopenshaw, SunCreator, Taborgate, Tanthalas39, Tarotcards, Technopat, Tentacle Monster, TheAlphaWolf, TimVickers, Time9, Titoxd, Tom harrison, Tomaxer, Tritium6, Twisted86, TwoOneTwo, Tyrenius, Uhh186, Ulgo, Unmake, Uogl, Usavi, User2004, UtherSRG, V111P, V8rik, Verdant04, Versus22, Viriditas, Vrenator, Wavelength, What a fat kid, Whosasking, Wikieditor06, Wikiklrsc, Wizardman, Wjejskenewr, WriterHound, XLerate, XP1, Xargque, Xasodfuih, Xbcj0843hck3, Xeaa, Xiuyechen, Yerpo, ZabMilenko, Zinzen, 518 anonymous edits

Louse *Source*: http://en.wikipedia.org/w/index.php?title=Louse *Contributors*: 16@r, 2over0, A8UDI, Aallyyssaa, Abigail-II, Acroterion, Adler777, Adzw101, Aimeef2, Alannah400197, Alansohn, AlexanderWinston, AlfredR, Anaxial, Andre Engels, AndrewHowse, Andy Marchbanks, Anna Lincoln, Anthony, Anthony Appleyard, Argosaurus, Arjun01, Art Carlson, Atif.t2, AureliaSylv, Barek-public, Bart133, Beetstra, Belinrahs, Biton, Blackicehorizon, Bluemoose, Bogey97, Bongwarrior, Bovineone, Bradeos Graphon, Brent513, Brentt, Brian0918, Bugboom, CDA, CactusCath, Camerong, CatherineMunro, Cerberus584, Cgoodwin, Chun-hian, Cletushenry, Corpx, Da Joe, DabMachine, Dadude3320, Darzbor, Deleteing, Deli nk, DerBorg, Djfrown, DocWatson42, Doctorfluffy, DoubleBlue, DragonflySixtyseven, Durga2112, Dylan anglada, E. Sn0 =31337=, Eastlaw, Eog1916, Epbr123, Esperant, Ethel Aardvark, Ezmindegy, Fallschirmjäger, FisherQueen, Foxjwill, Funkamatic, Furrykef, Gail, Gazz87, Gilliam, Gldren, Gogo Dodo, Gr1st, Gracenotes, Ground Zero, Gurch, H Debussy-Jones, Hasek is the best,

Hawaiian717, Hdt83, Headbomb, Hertz1888, Hit bull, win steak, Holliejade, Horatio, HuskyHuskie, Hydrogen Iodide, ILIKEPIEAWHOLEBUNCH, Iridescent, ItzMeAgain, Iysau, J.delanoy, JamesAM, Japanese Searobin, Jarebare, Jay Litman, Jj137, Jmh649, Jogloran, Jojhutton, Jrockley, Juansidious, Jusdafax, Kakofonous, Karategirl2010, KarlKarlson, Karthickbala, KateH, Katpatuka, Kbubbles, Kjoonlee, Kugamazog, Kukini, LOL, Lajos.rozsa, Lascorz, LeCire, Leahcim165, Leandrod, Lllamasrus, Lurazeke, MK8, MZaplotnik, Magioladitis, Mandarax, Maniac18, Marek69, MarkS, Martarius, Martin451, Mav, MaxSem, Mboverload, McDogm, Mentifisto, Mervyn, Michael Hardy, Mike Rosoft, Mileage, Mmm, Mnemo, MonkeyPieMan9, Morcheeba, Moshe Constantine Hassan Al-Silverburg, MrOllie, Muriel Gottrop, Mushroom, Musical Linguist, NPIC, Nakerlund, Nanodance, NawlinWiki, Neelix, Newyorkbrad, Nik42, Nixeagle, Nk, Nlu, Od Mishehu, Ohnoitsjamie, Omicronpersei8, OwenX, Patrick, Pavel Vozenilek, Plasticup, Pro bug catcher, Pscriba, QueenCake, RButcho, Raptur, Ratiocinate, Rchamberlain, Recognizance, Rees11, Riana, Rick lightburn, Rossami, Runwild2006, SG, Sawyermac, Scalimani, Seb az86556, Shawn2082, ShelfSkewed, Shelley Konk, Slgcat, Spasemunki, Spv926, Stemonitis, Stepheng3, Stwalkerster, Sutjo-18005, Svjetlana3, Syryll joy, T34, THEN WHO WAS PHONE?, Tangerines, Tangotango, Template namespace initialisation script, Tetracube, The Thing That Should Not Be, TheSimkin, Thebestofall007, Thenachster, Timwi, Tinton5, Tohd8BohaithuGh1, Tomaradze, Tomsefton, Totorotroll, Trfasulo, Tylerdmace, Ukexpat, Ursasapien, UtherSRG, Vanderesch, Vanished user 03, VernoWhitney, Vikte, Vkgfx, Waggers, Wapcaplet, WikHead, William Avery, Wlodzimierz, Wsiegmund, Xeaa, Yath, Yelocab, Yerpo, Zapvet, ZooFari, Zundark, آرمان, 335 anonymous edits

Mold *Source*: http://en.wikipedia.org/w/index.php?title=Mold *Contributors*: 04astoddart04, 12 Noon, 4twenty42o, A. Parrot, A8UDI, ABF, AK Auto, Abeg92, Addshore, AdjustShift, Adrian J. Hunter, Aesopos, Afluent Rider, Agent Smith (The Matrix), Ahoerstemeier, Airconswitch, Airplaneman, Aitias, Ajta, Aklibbee, Akronym, Alansohn, Alex.muller, Alex43223, Alexandrov, AlexiusHoratius, Allstarecho, Amux, Andre Engels, Andrew Kelly, Andries, Anetode, Anlace, Ann Stouter, Anonymi, Antandrus, Arcadian, Archanamiya, Artyom, Arwel Parry, Ascroggins, Ashman92, Atif.t2, Aude, Avicennasis, Avono, Avs5221, Azuizo, Badgernet, Basique, Bauer 1046, Bdowler, Beano, Bearcat, BeefRendang, Behaafarid, Bhazad, Biruitorul, Blanchardb, Blargmargsharg, Bob Blaylock, Bobo192, Bobzchemist, Bogey97, BorgHunter, BoundaryRider, Boy Cool67, Bped1985, Brabblebrex, Brainfsck, Brian ENVIRO, BrianY, Brumski, Bryan Derksen, Burnt*in*Effigy, Burntsauce, BurtAlert, Burzmali, Bzinga, CIreland, CMyVest, CO, CTwells, CWY2190, CWenger, Caichris, Caltas, Can't sleep, clown will eat me, CanadianCaesar, CanadianLinuxUser, Capricorn42, Cavfantic09, Ceramic2metal, Cflm001, Charitwo, Charles Nguyen, Charles Sturm, Charwinger21, Chasingsol, Chavster01, Chayama, Chongstar123, Chris G, Christopher Parham, Clicketyclack, CliffC, Codingmonkey, ColtM4, CommuterUser16, Cookie.cucumber, Corpx, Crash bandi, CrazyChemGuy, Crazycomputers, Creol, Crushthe, Cureden, CyrilleDunant, DARTH SIDIOUS 2, DJ Clayworth, DSRH, DVD R W, Dachshund, DanBealeCocks, Darth Panda, DaturaS, Davelane, Davewho2, Davewild, David Shear, DeadEyeArrow, Decltype, Deli nk, Delldot, DerHexer, DerWaschbar, Dev920, Dforest, Dina, Discospinster, DisillusionedBitterAndKnackered, Dlohcierekim, Dominus, Doodle77, Dorineforrest, Doubs24, Doulos Christos, Dr. Rondart, Dr.K., Dreadstar, Drmies, Dudydudeman, Dysepsion, Dysprosia, ESkog, East718, EdBever, EdEColbert, Edgar181, ElKevbo, Eliz81, Elkman, Emily4lyfe, Enviroboy, Enzo Aquarius, Epbr123, Erik9, Evan G, Evanreyes, Excirial, Fabrictramp, Falcon8765, Fastily, Fieldday-sunday, Fireworks, Flargman4, Flewis, Forluvoft, Free Bear, FrummerThanThou, Furrykef, Fuzzie, GB fan, GLaDOS, Gadfium, Gail, Gators222, Gdog10k, GeeJo, GeneralChan, Geniescience, GeorgeLouis, Georgechristodoulou, Gfoley4, Giants27, Giftlite, Gilliam, Ginkgo100, Giruvegan, Glane23, Glenn, Gludwiczak, Go Essendon, Gogo Dodo, GraemeL, Graibeard, Gramy, Greencircle, GregorB, Groundgamer, Gscshoyru, Guswandhi, Gzkn, Haham hanuka, Hammerhorn, Harriet horsfall, Hbent, Hdt83, Heimstern, Heron, Hibyespy, Hoo man, IRP, Illnab1024, IluvJoe, Immunize, Infocollider, Init, Inky, Inquisitus, Invitrovanitas, Iphone101, Iridescent, Irishguy, Ixfd64, J Milburn, J.delanoy, JDoorjam, JForget, JSarek, JSpung, Jake Wartenberg, Jamesontai, Jared Preston, Jatlas, Jcorkern, Jeff G., Jeltz, Jezebeau, Jfdwolff, JjusticeIV, JoJan, JohnOwens, Johnleemk, Jojit fb, Jonpro, Josh Grosse, Jplauriente, Juliancolton, Jusdafax, Jwoodger, KGasso, KJS77, Kaol, Karenjc, Kayleeroxx, Keegan, Keilana, Kembangraps, Kenny44, Kgasc006, Kidmilo10, Killiondude, Kilva, Kimyko, Kiniki10, Kittykatt4134, Konman72, KonradWallenrod, Kubigula, Kvn8907, Kylebeezy0, L Kensington, LOL, La Pianista, LeaveSleaves, Legion fi, Leuko, LizardJr8, Luckyherb, MBob, MER-C, MPF, Makemi, Malo, Marc Venot, MarcoTolo, Marek69, MarkSutton, Marnanel, Martpol, Materialscientist, Mato, Maxis ftw, McSly, Meaningful Username, Menchi, Mentifisto, Mexican Clayton, Mike Rosoft, Mike6271, Milleb42, Milosh1414, Mkweise, Mnh, Mo0, Mod.torrentrealm, Mononomic, Moreschi, MortimerCat, Mouldyalmy321, MrDarwin, Ms.Reagers, Multimotyl, Mwanner, My76Strat, Mysdaao, N3362, N419BH, NPIC, Nabarry, Nakon, NawlinWiki, Neilc, Nemetona, Neoballmon, Neocarpetbagger, Netkinetic, Neurolysis, Ngebendi, Nicholas simmons, Nick C, Nixdorf, NixonB, Northumbrian, NotFromUtrecht, Octahedron80, Odie5533, OffsBlink, Ohnoitsjamie, Oknedar, OlEnglish, Omicronpersei8, Once in a Blue Moon, Optichan, Orange Suede Sofa, Oxymoron83, Pablo442, Panscient, Pattybutch, Pdery, Pentasyllabic, Peripitus, Philip Trueman, Pi, Pieman9712, PierreAbbat, Pigsonthewing, Pinethicket, Pingveno, Pinkadelica, Pixel8, Pledger166, Poppy1128, PrestonH, PrettyCoCo818, PrincessofLlyr, Pro crast in a tor, Promethean, Pteron, Puchiko, Pupster21, Purplemagik, QuadrivialMind, Qxz, R, RL0919, Radbug, Raproducer, Reach Out to the Truth, Reaper Eternal, Remodeling101, Repliedthemockturtle, Restorationsos, Rettetast, RevMK, RexNL, Rgrof, RipperWest, Rj4179, Rmhermen, Roboctopus, Rocket71048576, Roger McLassus, Romanskolduns, Ronhjones, Ronz, Rrburke, SEWilco, Saperaud, Sasuke487, Scarian, Sceptre, Scharks, Schuym1, Scimitar, Scwlong, Senator Palpatine, Sergio sanchez, Seth Whales, Shadeybaby, Shadowjams, ShampooCell, Shanel, Shawnd64, Sheila del Barrio, Sheogorath, Shoeofdeath, Simplechic, Sir Vicious, Sit277, Skater, Skull150, Skullblast, SlimVirgin, Slowking Man, Slysplace, Snowmanradio, Snowolf, Snufflesowner, Sonett72, Spasemunki, Spick And Span, Spitfire ch, SpuriousQ, Spydercanopus, SquidThing, Sten, Stephenb, Steven Weston, Steven Zhang, Stwalkerster, SuperHamster, Sven Manguard, Svetovid, Switchercat, Szyslak, THEN WHO WAS PHONE?, TOF1000, Tapir Terrific, TastyPoutine, TedE, Texasbacon, Thatguyflint, The High Fin Sperm Whale, The Obfuscator, The Rambling Man, The Thing That Should Not Be, Theforager, Themfromspace, Thingg, Tide rolls, TigerShark, Tim Starling, Tohd8BohaithuGh1, Tommy2010, Travelbird, Tresiden, Trusilver, Tuktusong, TulaneGrad, Tumadoireacht, Twentyquid, Ueberzahl, UriahHeepster33220, Utcursch, Valerielgreen, Valermos, Vary, Verne Equinox, Versus22, Vrenator, WadeSimMiser, Wallsts, Walor, Warofdreams, Wereon, West.andrew.g, WhatamIdoing, WikidSmaht, Wikipelli, William Avery, Wimox, Wizard191, Wmcarlson, Wolfrock, Worksafe, Wysprgr2005, Xiahou, Xioxox, Xp54321, Xytor500, Yamamoto Ichiro, Zambani, Zarggg, Zargos10, Zebruh, Zhou Yu, Zico24, Zipzip50, Zokou, Zondor, Zudduz, 1410 anonymous edits

Fungus *Source*: http://en.wikipedia.org/w/index.php?title=Fungus *Contributors*: -Ilhador-, 144.92.164.xxx, 24.108.233.xxx, AC+79 3888, ARog, Aaron Schulz, Abductive, Acalamari, Adam78, Adambro, Adamcisme123, Adashiel, Addihockey10, Adrian J. Hunter, AdultSwim, Agathman, Ahoerstemeier, Aitias, Ajaxkroon, Alan Liefting, Alan Rockefeller, Alansohn, Alanyates26, Alexei666, AlexiusHoratius, Alksentrs, Alphachimp, AmericanXplorer13, AnakngAraw, Anclation, Andre Engels, Andres, Andrew512, Andrewlp1991, Andux, Andycjp, Angel Shockers, Angela, Aniten21, Annabel, Antandrus, Anthony Appleyard, Anypodetos, Apollonius 1236, Aquatics, Arcade777, Arcadian, Art LaPella, Arthena, Artichoker, Ashishbhatnagar72, Ashmoo, Astronautics, Atlant, AvicAWB, Awadewit, AxelBoldt, Axl, AzaToth, Azhyd, Badagnani, Badseed, Banana04131, Barfooz, Bart133, Batmaninbandages, Bayle Shanks, Bballoakie, Bcorr, Beetstra, Ben DeRoy, Bender235, Bendzh, Bggoldie, Big Bird, Bigdkbrew, Bill37212, Blah28948, BlueCanoe, Bmicomp, Bobblehead, Bobblewik, Bobo192, Bomac, Bongwarrior, Booksworm, BorgQueen, Bradley Grillo, Brandonrush, Brewhaha@edmc.net, Briséis, BrownHairedGirl, Bruce Marlin, Brya, Btboy500, Bunchofgrapes, CDN99, CIreland, Calair, CaliforniaAliBaba, Caltas, Calwiki, Calz22123, CambridgeBayWeather, Canjth, CardinalDan, Carrp, Casliber, Cayte, Cdlw93, Ceramic2metal, Cerejota, Ceyockey, Chanting Fox, Chaos, Chatool, Chazchaz101, Chrislk02, Christophergomez, Chun-hian, Citizen Premier, Claire van der Meer, Clicketyclack, CommonsDelinker, Conversion script, Corryong, Cpink, Crampedson, Crazy rollz, Cremepuff222, Crevox, Crusadeonilliteracy, Cryptic C62, CryptoDerk, Cucamber, Cundi, Cvstevani, DGG, DVD R W, DabMachine, Dabomb87, Dacoutts, Dajhorn, Dana boomer, Daniel11, Darcrist, David D., DavidMcCabe, Daytondavis, Dbenbenn, Ddusenbery, Debivort, December21st2012Freak, Denger, Der Eberswalder, DerHexer, Derek Andrews, Deville, Dinoguy2, DirkvdM, Djmckee1, Dlloyd, Dlohcierekim's sock, DonDaMon, Donald Albury, Doulos Christos, Download, Dragonnas, Drenaline, Drgnu23, Drunkenmonkey, Dysepsion, E0steven, EDUCA33E, ENeville, ESkog, Edward321, Eequor, Eevans12392, EliasAlucard, Eluchil404, Emperorbma, EncycloPetey, Eras-mus, Eraserhead1, Eric119, Escape Orbit, Espoo, Ettrig, Everard Proudfoot, Everyking, EvocativeIntrigue, Ewlyahoocom, Excirial, Extransit, Ezhuttukari, FF2010, Falcon8765, Fanghong, FastLizard4, Favonian, Fawcett5, Felizdenovo, Feydey, Fieldday-sunday, Foobar, Fox2030, Frankman28, Freakofnurture, FreplySpang, Frosted14, Funnybunny, Furrykef, GVnayR, Gabbe, Gadfium, Galloramenu, Galorr, Galoubet, Gbalitgs, Gbleem, Gdarin, Gdr, GeeJo, Gerpictus, Gianfranco, Giftlite, Gilliam, Glenn, Gogo Dodo, Goobergunch, Graham87, GrahamColm, Graminophile, Grayshi, GreenReaper, Grosscha, Grundig, Gtg204y, Guanaco, Gunmetal Angel, Gurko, Gwendal, Gwhodgson, HANDICAPPEDhero, Hadal, Haha12345ownd, Haham hanuka, HalfShadow, Hardyplants, Harraser, Hayley Tales, Hc.shruthi, Heah, Helikophis, Hellbus, HenkvD, HereToHelp, Heron, Heungki1234, Hi bye man, Hike395, Hippophaë, HunterZ, Hurricane111, Husond, Hut 8.5, Hydroxonium, ImperfectlyInformed, IndyLawSteve, Into The Fray, Irandegan, Irate Vermin, Irishguy, IronChris, Ixfd64, J Milburn, J. Finkelstein, J. Spencer, J.delanoy, J04n, JDG, JForget, JHunterJ, JLaTondre, Jackol, Jacobolus, Jaknouse, Jallan, JamesHoadley, Jamie487833, Jatlas, Jatlas2, Jay2332, Jchernandez, Jdavidb, Jdawg4820, Jedi Grzyb, Jefffire, JemGage, JerrySteal, Jitterro, Jjamison, Jmeppley, Jml112, JoJan, JoanneB, Joaotg, John254, JohnOwens, Johndoh75, Jose Ramos, Josh Grosse, Josh the Nerd, JosiahHenderson, Joyous!, Jsbillings, Jsky1985, Junglecat, Jusdafax, Just Another Dan, Jvbishop, Jwillbur, Jwissick, K3beast, Kaini, Kansan, Kappa, Karen Johnson, Karkiked, Karl-Henner, Katieh5584, Kaushal mehta, Kcamnairb, Kelovy, Kesh, Kevin, Kibanc, Killdevil, Kingturtle, Kipala, Kirill Lokshin, Kjaer, Kku, Knakts, KnowledgeRequire, Knutux, Koppas, Kostya, Kralizec!, Kuru, Kwamikagami, L Kensington, Labelledamesansmerci, Ladida95, Lady of the shadows, Lambo911, Lambyte, Lando Calrissian, Lankiveil, Lavateraguy, Lcol01, Lectonar, Leebigmanshe, Lesfreck, Lesgles, Levineps, Lexor, Lightmouse, LiquidGhoul, LizardJr8, Llort, Llull, LooopayyyyyLunnnnarrrr, Los editadores, Lotusduck, Luehea, Luigifan, Luigizanasi, Luk, Lumos3, MZMcBride, Macronyx, Madpassion, Magnus Manske, Mailer diablo, Makemi, Malljaja, Mandarax, ManosGR, Marek69, MarkSutton, Marshman, Master of Puppets, Masterjamie, Masterofsuspense2, Materialscientist, Mattarata, Matthew Wilson, Mattisse, Mav, Max Naylor, Mbr1983, Menchi, Mentifisto, Merope, Mgiganteus1, Mhking, Michael Devore, MidgleyDJ, Mike Young, Mike2vil, MikeAus, Milton Stanley, Minghong, Minisconom, Mishalak, Moeron, Moon&Nature, Moreschi, Motyka, Mporemba, Murphyen, Mushin, Mustafaa, Mycota, Nagec, Nakon, Nandesuka, Narutodude000, NatureA16, NawlinWiki, Nbarth, NebuchadnezzarN, Neelix, Neil916, Nethgirb, NewEnglandYankee, Nibuod, Nick683, Nihiltres, NoahElhardt, NotAnonymous0, Nothlit, Notnihekaj, NuclearWarfare, Nuggetboy, OOODDD, Oatmeal batman, Obli, Ocaasi, Ocatecir, OhanaUnited, Olli, Onco p53, Onorem, Onrole, Orangemarlin, Otlw112, Ottawa4ever, Ouishoebean, Oxymoron83, PDH, PanAndScan, Pauli133, Pb30, Pdcook, Pdculbert, Pekinensis, Peripitus, Perry04061994, Peruvianllama, Peter G Werner, Phil Boswell, Phileas, Phyzome, Pietro6, Pigsonthewing, Piledhigheranddeeper, Pinethicket, Pinky sl, Pippu d'Angelo, Plasticup, Plindenbaum, Poeloq, Poketaru Man, Preacherdoc, Proofreader77, Pseudo-Richard, PseudoOne, Puffball, Pupster21, Pzrmd, Qatter, Quercusrobur, Qxz, R0uge, RJASE1, Radishes, Raelx, Rafti Institute, Railfence, Rajasingh, Randyoo, Rasenganz, Ravidreams, Reddi, Rich Farmbrough, Richard001, Richhoncho, Ringler, Rjwilmsi, Rmx256, RobertG, Ronz, RoryReloaded, Roxwhal, Roy Bateman, RoyBoy, Royboycrashfan, Rreagan007, Ruud Koot, Ryan the asomly cool man, Ryulong, S Marshall, Sade, Saebjorn, Sahilbarua, Sam Korn, Samsara, Sandstein, SandyGeorgia, Santigore, Sasata, Scarian, Scharks, Scoo, Scottalter, Sean gorter, SeanMD80, SeanMack, Seb az86556, Seb951, Seglea, Sengkang, Sergio sanchez, Shanel, Shoessss, Silence, Silik, SilkTork, SiobhanHansa, SirIsaacBrock, Sjakkalle, Sjones23, Skizzik, Sky Attacker, Slecki19, Smack, Smith609, Smooth O, Snigbrook, Snowmanradio, Sohmc, Someone else, SpLoT, Spaceace922, Spe0059, Speight, Spellmaster, Spidermonkey, Spitfire ch, SpuriousQ, Starvinsky, SteinbDJ, Stemonitis, Stephen G. Brown, Stephen MUFC, Stephenb, Stepp-Wulf, Steve3849, Stevenrasnick, Stevertigo, Stevey7788, Strobilomyces, StuffOfInterest, Summerbell, Summerbellrc, Swedish fusilier, Sylent, Sylverfysh, TUF-KAT, Tariqabjotu, Tbhotch, TedPavlic, Teebol, Temuler, Teply, Terence, TestPilot, The Thing That Should Not Be, TheAlphaWolf, TheKMan, TheYoungerLohan, Theflea200, Thespian, Tide rolls, Tim Q. Wells, TimVickers, Timir2, Timo Honkasalo, Todfox, Tom Lougheed, Tomiko72, Tony Fox, TotoBaggins, Touchstone42, Tree Biting Conspiracy, Tree1983, Troop350, Turpin, U.S.A.U.S.A.U.S.A., Ucanlookitup, User27091, Userafw, UtherSRG, Vanderesch, Varden12, Velho, Velosa, Vignaux, Viktor-viking, Vina, Vipinhari, Viriditas, VonKvidtbaen, Vsmith, WJBscribe, Waggers, Watcharakorn, Wavelength, Wbrameld, Wedian, Werothegreat, Wesley J M, Wetman, Wham Bam Rock II, WhatamIdoing, WiiWillieWiki, Wiki wiki1, Wikipe-tan, Wikiskimmer, Wilindgren, WillMak050389, William Avery,

Willtron, Wimt, Wj32, Wknight94, Wlodzimierz, Wmahan, Wolfdog, Wood Thrush, WormRunner, Wowbobwow12, Xanzzibar, Xenium, Xeno, Xmnemonic, Yamamoto Ichiro, Yidisheryid, YixilTesiphon, Ynhockey, Zachorious, ZimZalaBim, Zsinj, Zundark, Ô, 1034 anonymous edits

Tick *Source*: http://en.wikipedia.org/w/index.php?title=Tick *Contributors*: (aeropagitica), 10metreh, A little insignificant, A. B., A. Parrot, AThing, AVand, Abanima, Abductive, Acetic Acid, Affe, Ageoflo, Ahoerstemeier, Aitias, Aka, Ale jrb, AlexPlank, AlexiusHoratius, Allison Stillwell, Altenmann, AndrewHowse, Andrewpmk, Antandrus, Antics05, Antonio Lopez, Apokryltaros, Arachn0, Arcadian, ArchonMagnus, Arsine, Arto.kem, AshLin, Audiosmurf, Audriusa, Avitya, Awien, Bartosik, Beano, Beastiepaws, Belovedfreak, Benajnim, Bkonrad, Blueboxz, Bob Burkhardt, Bob98133, Bobdoe, Bobo192, Bogdangiusca, BoundaryRider, Bowfee, Braddirt, Bruce Marlin, Bumbulski, Byzz, CIS, CaliforniaLyme, CamJM, Can't sleep, clown will eat me, CanOfWorms, CanadianLinuxUser, Caulde, Cgoodwin, Chamaeleon, Christopher denman, Christopherjfoster, Chriswaterguy, Chuunen Baka, Chzz, CiudadanoGlobal, Coasterlover1994, Cocytus, Codieraji, Cohee, CommonsDelinker, ConCompS, Connecticuttickcontrol, Conscious, Conti, Cortlan, Cpl Syx, CrakerFang, Creidieki, Cryonic07, Cult2y, CuteHappyBrute, D o m e, DARTH SIDIOUS 2, DLew978, Dachshund, Daniel.Cardenas, Dante Alighieri, Dark jedi requiem, Daven200520, David Koller, David.Monniaux, Derekduke77, Derlay, Dew Kane, Diderot, Diego Grez, Discospinster, Dlae, Doctorfluffy, Docvee, Dodo bird, Dontaskme, DreamGuy, Dumbo12, Dysmorodrepanis, ESkog, Ecarlson81, Eclipsed Moon, Edward, Eeekster, Elektron, Elf, Encephalon, Epbr123, Eras-mus, Erohov, Espnpsycho, Etxrge, Eurosong, Evercat, Everyking, Excirial, Falcon8765, Fenevad, Ferdinand Pienaar, Frazpaterson, Fumitol, Funandtrvl, Furrykef, Fusionmix, Fæ, Garion96, Gary Alpert, Gary King, Georgette2, Ghouse, Gilliam, Goblin, Gouerouz, Grafikm fr, GregorB, Ground Zero, Guiltyspark, Gyrofrog, Haham hanuka, HamburgerRadio, Hardisk44, Harryboyles, Haukurth, Haymaker, Heado, Hephaestos, Hereforhomework2, Herroyalmaj, Hertz1888, Howcheng, Huw Powell, Hydrargyrum, II MusLiM HyBRiD II, Iain marcuson, Innotata, Iridescent, IronChris, Iviney, J Hazard, J.delanoy, JForget, JMK, Jaan513, Jacques Brando, Jary, Jddriessen, Jeremykemp, Jim10701, Jimworm, Jklsc, Jmagoon, Joewithajay, Jojhutton, Jonathan W, Josh Grosse, Jotunn, Jp498, Jpgordon, Jrtayloriv, Jtherlow, Julian Mendez, Jusdafax, Justallofthem, Jyhcarie, K10wnsta, KCavanaugh, Kaaos, Kaarel, Karaa91, Kategriff, KathrynLybarger, Keelm, Keilana, Kembangraps, Kevin X PeePee, Kevmin, Khargas, Kinickie, KnowledgeLord, KrakatoaKatie, Kralizec!, Kummi, Kurtan, Kutera Genesis, La Pianista, Lachatdelarue, Lamiot, Lazylaces, Lebhill, Leon7, Liftarn, Lighthead, Lisapollison, Lotje, Luckyz21, Luvitsa, Lyinginbedmon, Macaddct1984, Magister Mathematicae, MagneticFlux, Magnus Manske, Maheshkumaryadav, Mandarax, Marc Venot, MarkMentovai, Martino3, Matilda, Mattachu, Maxyman, Meaghan, Mecil, Melnuk12, MichaelGood, Michele123, MikeUMA, Mikedeprez, Minesweeper, Monty845, Moogwrench, Mr. Scholarly Guy, Mrmiscellanious, NPIC, Nakon, NaliniL, NawlinWiki, Nemoiscool, Neparis, Nepenthes, NessieVL, Neurovelho, NewEnglandYankee, NielsenGW, Nivix, Northumbrian, Notafly, Nyttend, O.anatinus, Obersachse, Ocee, Oda Mari, Ohnoitsjamie, OlEnglish, Old Moonraker, Orangutan, PRB, Palea, PanAndScan, Pandora, ParticleMan, Patrick, Pekinensis, PhJ, Phasespace, Philip Trueman, PhoenixSunburn, Piano non troppo, Pinethicket, Pizza Puzzle, Planckton, Potatoswatter, PotentialDanger, Pragmatist, Prashanthns, Promethean, ProteinGod, Purgatory Fubar, Quebec99, RTC, Raul654, Reed.lacy, Rehmeyer, RepublicanJacobite, Res2216firestar, Rex the first, RexNL, Richhoncho, Richwales, Rigby27, Rklawton, Rolypolyman, Rsabbatini, Rubberducky024, Ruby miette, Runefrost, Runwild2006, RyanCross, SJP, Sam Korn, Sapphirine, Sarefo, Sarsaparilla, Scarian, SchmuckyTheCat, Sciurinæ, Scott Johnson, SeLfkiLL, Sean D Martin, Search4Lancer, Seb az86556, Seraphim, Shadowjams, Shall I Make Wikipedia A Boom Town?, Shawisland, Shenme, Signalhead, SimDarthMaul, Sinuhe, SlateGrey, Smooth O, Smooth0707, Snailwalker, Snowmanradio, Sobreira, Sonett72, SpaceFlight89, Specialkay985, Spitfire, Spydercanopus, Squamate, Stemonitis, SteveBaker, Steven Weston, Stone, Tahdah, Tannin, Technel, The Duke of Waltham, The High Fin Sperm Whale, The Thing That Should Not Be, Theanthrope, Thingg, Thomas.Hedden, Tickbuddy, Ticklemygrits, Tide rolls, Timo Honkasalo, Tiptopper, Tnd, Tom, Tombomp, Tommy2010, TotoBaggins, Toyokuni3, Trees123, Trfasulo, Truehistoryjvba, Tsgttucker, Ulric1313, Uncle Milty, Unyoyega, UpstateNYer, Utcursch, Valerius Tygart, VasilievVV, Vectorborne, Vianello, Violask81976, Voyagerfan5761, Vrenator, Vsundar, Vulturelainen, WadeSimMiser, Warszawa1000, WikiLaurent, WikipedianMarlith, Wisdom89, Wknight94, Woohookitty, Worldwalker, Xabian40409, Yamamoto Ichiro, Yerpo, Yesuto15, Ynbff97, ZeroOne, °, Саша Стефановић, احمدغامدي.24, ماني, , 745 anonymous edits

Image Sources, Licenses and Contributors

file:Drawing-1.png *Source*: http://en.wikipedia.org/w/index.php?title=File:Drawing-1.png *License*: Creative Commons Attribution-Sharealike 3.0 *Contributors*: User:Bugboy52.40

File:Loudspeaker.svg *Source*: http://en.wikipedia.org/w/index.php?title=File:Loudspeaker.svg *License*: Public Domain *Contributors*: Bayo, Gmaxwell, Husky, Iamunknown, Myself488, Nethac DIU, Omegatron, Rocket000, The Evil IP address, Wouterhagens, 9 anonymous edits

File:Fiddler beetle morphology diagram.svg *Source*: http://en.wikipedia.org/w/index.php?title=File:Fiddler_beetle_morphology_diagram.svg *License*: Creative Commons Attribution-Sharealike 3.0 *Contributors*: User:Bugboy52.40

File:Beetle takeoff.jpg *Source*: http://en.wikipedia.org/w/index.php?title=File:Beetle_takeoff.jpg *License*: Creative Commons Zero *Contributors*: User:Kpts44

File:Scarabaeus viettei head.jpg *Source*: http://en.wikipedia.org/w/index.php?title=File:Scarabaeus_viettei_head.jpg *License*: GNU Free Documentation License *Contributors*: User:AxelStrauss, user:bugboy52.40

File:Cephalota circumdata circumdata front.jpg *Source*: http://en.wikipedia.org/w/index.php?title=File:Cephalota_circumdata_circumdata_front.jpg *License*: Creative Commons Attribution-Sharealike 3.0 *Contributors*: User:Siga

File:Acilius sulcatus.gemeiner furchenschwimmer.jpg *Source*: http://en.wikipedia.org/w/index.php?title=File:Acilius_sulcatus.gemeiner_furchenschwimmer.jpg *License*: Public Domain *Contributors*: Kai Kluwe

File:Internal morphology of Coleoptera.svg *Source*: http://en.wikipedia.org/w/index.php?title=File:Internal_morphology_of_Coleoptera.svg *License*: Creative Commons Attribution-Sharealike 3.0 *Contributors*: User:Bugboy52.40

File:Punctate Flower Chafer.jpg *Source*: http://en.wikipedia.org/w/index.php?title=File:Punctate_Flower_Chafer.jpg *License*: unknown *Contributors*: Calliopejen, Dcoetzee, Fir0002, Kersti Nebelsiek, Rocket000, Sarefo

File:Life cycle of stag beetle.svg *Source*: http://en.wikipedia.org/w/index.php?title=File:Life_cycle_of_stag_beetle.svg *License*: Creative Commons Attribution-Sharealike 3.0 *Contributors*: User:Bugboy52.40

Image:Curl grub.jpg *Source*: http://en.wikipedia.org/w/index.php?title=File:Curl_grub.jpg *License*: Creative Commons Attribution-Sharealike 3.0 *Contributors*: User:99of9

File:Photinus pyralis Firefly 2.jpg *Source*: http://en.wikipedia.org/w/index.php?title=File:Photinus_pyralis_Firefly_2.jpg *License*: Creative Commons Attribution-Sharealike 2.0 *Contributors*: art farmer from evansville indiana, usa

File:Scarabaeus laticollis.jpg *Source*: http://en.wikipedia.org/w/index.php?title=File:Scarabaeus_laticollis.jpg *License*: Creative Commons Attribution 2.5 *Contributors*: Rafael Brix

File:Orange Blister Beetle (Mylabris pustulata) on Ipomoea carnea W IMG 0593.jpg *Source*: http://en.wikipedia.org/w/index.php?title=File:Orange_Blister_Beetle_(Mylabris_pustulata)_on_Ipomoea_carnea_W_IMG_0593.jpg *License*: Creative Commons Attribution 3.0 *Contributors*: User:J.M.Garg

File:Pegesimallus sp robberfly.jpg *Source*: http://en.wikipedia.org/w/index.php?title=File:Pegesimallus_sp_robberfly.jpg *License*: unknown *Contributors*: Chmee2, Dysmachus, Dysmorodrepanis, H005, Herald Alberich, Kersti Nebelsiek, Kevmin, Muhammad Mahdi Karim, Slashme

File:Gemeiner Widderbock 4483.JPG *Source*: http://en.wikipedia.org/w/index.php?title=File:Gemeiner_Widderbock_4483.JPG *License*: Creative Commons Attribution-Sharealike 3.0 *Contributors*: User:Hedwig Storch

File:Ambrosia beetle life cycle.svg *Source*: http://en.wikipedia.org/w/index.php?title=File:Ambrosia_beetle_life_cycle.svg *License*: Creative Commons Attribution-Sharealike 3.0 *Contributors*: User:Bugboy52.40

File:Baltic amber Coleoptera Scraptiidae.JPG *Source*: http://en.wikipedia.org/w/index.php?title=File:Baltic_amber_Coleoptera_Scraptiidae.JPG *License*: Attribution *Contributors*: Anders L. Damgaard - www.amber-inclusions.dk

File:Prachtkäfer aus der Grube Messel.JPG *Source*: http://en.wikipedia.org/w/index.php?title=File:Prachtkäfer_aus_der_Grube_Messel.JPG *License*: GNU Free Documentation License *Contributors*: Original uploader was Gerbil at de.wikipedia

File:Phylogenetic chart of Coleoptera.svg *Source*: http://en.wikipedia.org/w/index.php?title=File:Phylogenetic_chart_of_Coleoptera.svg *License*: Creative Commons Attribution-Sharealike 3.0 *Contributors*: User:Bugboy52.40

File:Cotton_boll_weevil.jpg *Source*: http://en.wikipedia.org/w/index.php?title=File:Cotton_boll_weevil.jpg *License*: Public Domain *Contributors*: Original uploader was Ellmist at en.wikipedia

File:Potato beetle larvae.jpg *Source*: http://en.wikipedia.org/w/index.php?title=File:Potato_beetle_larvae.jpg *License*: GNU Free Documentation License *Contributors*: ComputerHotline, MPF, Quadell, Ranveig, Xvazquez

File:Lady-beetle-close-up.jpg *Source*: http://en.wikipedia.org/w/index.php?title=File:Lady-beetle-close-up.jpg *License*: Creative Commons Attribution-Sharealike 2.5 *Contributors*: Thomas G. Moertel (Tmoertel at en.wikipedia)

Image:Mealworm 01 Pengo.jpg *Source*: http://en.wikipedia.org/w/index.php?title=File:Mealworm_01_Pengo.jpg *License*: Creative Commons Attribution-Sharealike 3.0 *Contributors*: User:Pengo

File:Zopheridae_jewelry_sjh.jpg *Source*: http://en.wikipedia.org/w/index.php?title=File:Zopheridae_jewelry_sjh.jpg *License*: Creative Commons Attribution-Sharealike 2.5 *Contributors*: Dysmorodrepanis, Kugamazog

Image:Egyptian Dung beetle2008.jpg *Source*: http://en.wikipedia.org/w/index.php?title=File:Egyptian_Dung_beetle2008.jpg *License*: GNU Free Documentation License *Contributors*: Ryan Postlethwaite Original uploader was Ryan Postlethwaite at en.wikipedia

Image:Egypt.KV6.04.jpg *Source*: http://en.wikipedia.org/w/index.php?title=File:Egypt.KV6.04.jpg *License*: GNU Free Documentation License *Contributors*: Aavindraa, Aoineko, Gospodar svemira, Hajor, JMCC1, Juiced lemon, Marcus Cyron, Mmcannis, Neithsabes

file:Oryzaephilus.jpg *Source*: http://en.wikipedia.org/w/index.php?title=File:Oryzaephilus.jpg *License*: Public Domain *Contributors*: Original uploader was Andybrookestar at en.wikipedia

file:Scanning_Electron_Micrograph_of_a_Flea.jpg *Source*: http://en.wikipedia.org/w/index.php?title=File:Scanning_Electron_Micrograph_of_a_Flea.jpg *License*: Public Domain *Contributors*: Content Provider(s): Centers for Disease Control and Prevention (CDC) / Janice Carr

File:Scheme flea anatomy-en.svg *Source*: http://en.wikipedia.org/w/index.php?title=File:Scheme_flea_anatomy-en.svg *License*: Creative Commons Attribution 3.0 *Contributors*: User:Al2

File:HookeFlea01.jpg *Source*: http://en.wikipedia.org/w/index.php?title=File:HookeFlea01.jpg *License*: Public Domain *Contributors*: Robert Hooke

File:Flea Larva.jpg *Source*: http://en.wikipedia.org/w/index.php?title=File:Flea_Larva.jpg *License*: GNU Free Documentation License *Contributors*: User:Kalumet

File:Macskabolha.png *Source*: http://en.wikipedia.org/w/index.php?title=File:Macskabolha.png *License*: GNU Free Documentation License *Contributors*: User:Outesticide

File:Fleabite.JPG *Source*: http://en.wikipedia.org/w/index.php?title=File:Fleabite.JPG *License*: unknown *Contributors*: Yerpo, 1 anonymous edits

File:Fleabite-closeup.jpg *Source*: http://en.wikipedia.org/w/index.php?title=File:Fleabite-closeup.jpg *License*: Public Domain *Contributors*: Bootlog

File:2008-08-06 Flea and tick powder applied to White German Shepherd pup.jpg *Source*: http://en.wikipedia.org/w/index.php?title=File:2008-08-06_Flea_and_tick_powder_applied_to_White_German_Shepherd_pup.jpg *License*: GNU Free Documentation License *Contributors*: User:Specious

File:Fleadirt.jpg *Source*: http://en.wikipedia.org/w/index.php?title=File:Fleadirt.jpg *License*: unknown *Contributors*: Original uploader was Dr Zak at en.wikipedia

file:Drosophila melanogaster - side (aka).jpg *Source*: http://en.wikipedia.org/w/index.php?title=File:Drosophila_melanogaster_-_side_(aka).jpg *License*: Creative Commons Attribution-Sharealike 2.5 *Contributors*: user:Aka

File:55542main maflies med.jpg *Source*: http://en.wikipedia.org/w/index.php?title=File:55542main_maflies_med.jpg *License*: Public Domain *Contributors*: Dysmachus, Sarefo, Superborsuk, Wst

File:Drosophila egg.png *Source*: http://en.wikipedia.org/w/index.php?title=File:Drosophila_egg.png *License*: GNU Free Documentation License *Contributors*: User JWSchmidt on en.wikipedia

File:Fruit flies.jpg *Source*: http://en.wikipedia.org/w/index.php?title=File:Fruit_flies.jpg *License*: Creative Commons Attribution 3.0 *Contributors*: User:TheAlphaWolf

File:Drosophila melanogaster lab cultures.jpg *Source*: http://en.wikipedia.org/w/index.php?title=File:Drosophila_melanogaster_lab_cultures.jpg *License*: Creative Commons Attribution 3.0 *Contributors*: User:Trick17

File:EyeColors.jpg *Source*: http://en.wikipedia.org/w/index.php?title=File:EyeColors.jpg *License*: Public Domain *Contributors*: Kersti Nebelsiek, Ktbn, Marcok, 2 anonymous edits

File:Drosophila-chromosome-diagram.jpg *Source*: http://en.wikipedia.org/w/index.php?title=File:Drosophila-chromosome-diagram.jpg *License*: Public Domain *Contributors*: Steven J. Baskauf

File:Drosophila m oogenesis.png *Source*: http://en.wikipedia.org/w/index.php?title=File:Drosophila_m_oogenesis.png *License*: GNU Free Documentation License *Contributors*: Dysmachus, Mindmatrix, Superborsuk

File:Fly eye stereo pair.png *Source*: http://en.wikipedia.org/w/index.php?title=File:Fly_eye_stereo_pair.png *License*: GNU Free Documentation License *Contributors*: Carnildo, Chowbok, Dysmachus, Kersti Nebelsiek, Mindmatrix, Superborsuk

File:Rh1.jpg *Source*: http://en.wikipedia.org/w/index.php?title=File:Rh1.jpg *License*: Public Domain *Contributors*: Hokiedokie

file:Fahrenholzia pinnata.JPG *Source*: http://en.wikipedia.org/w/index.php?title=File:Fahrenholzia_pinnata.JPG *License*: unknown *Contributors*: photo by Vincent S. Smith

File:Ricinus bombycillae (Denny, 1842).JPG *Source*: http://en.wikipedia.org/w/index.php?title=File:Ricinus_bombycillae_(Denny,_1842).JPG *License*: Public Domain *Contributors*: User:Lajos.Rozsa

File:Trinoton anserinum (Fabricius, 1805).JPG *Source*: http://en.wikipedia.org/w/index.php?title=File:Trinoton_anserinum_(Fabricius,_1805).JPG *License*: Public Domain *Contributors*: User:Lajos.Rozsa

File:Lice image01.jpg *Source*: http://en.wikipedia.org/w/index.php?title=File:Lice_image01.jpg *License*: GNU Free Documentation License *Contributors*: Dysmorodrepanis, Luis Fernández García

File:Louse_diagram,_Micrographia,_Robert_Hooke,_1667.jpg *Source*: http://en.wikipedia.org/w/index.php?title=File:Louse_diagram,_Micrographia,_Robert_Hooke,_1667.jpg *License*: Public Domain *Contributors*: Brian0918, Kilom691, MushiHoshiIshi

File:Mouldy Clementine.jpg *Source*: http://en.wikipedia.org/w/index.php?title=File:Mouldy_Clementine.jpg *License*: Creative Commons Attribution-Sharealike 3.0 *Contributors*: NotFromUtrecht

Image:Moldy Bread.jpg *Source*: http://en.wikipedia.org/w/index.php?title=File:Moldy_Bread.jpg *License*: Creative Commons Attribution 3.0 *Contributors*: User:Once in a Blue Moon

Image:DecayingPeachSmall.gif *Source*: http://en.wikipedia.org/w/index.php?title=File:DecayingPeachSmall.gif *License*: Creative Commons Attribution-Sharealike 2.0 *Contributors*: Andrew Dunn (User:Solipsist)

Image:Moldy nectarines.jpg *Source*: http://en.wikipedia.org/w/index.php?title=File:Moldy_nectarines.jpg *License*: GNU Free Documentation License *Contributors*: 790, ComputerHotline, Roger McLassus, 3 anonymous edits

Image:DSC01904.JPG *Source*: http://en.wikipedia.org/w/index.php?title=File:DSC01904.JPG *License*: Creative Commons Attribution 3.0 *Contributors*: User:Jplauriente

Image:Moldy Strawberries.jpg *Source*: http://en.wikipedia.org/w/index.php?title=File:Moldy_Strawberries.jpg *License*: Creative Commons Attribution-Sharealike 3.0 *Contributors*: User:Kevin_Payravi

Image:20091129_210544_FungusFromContourCabinAirFilter.jpg *Source*: http://en.wikipedia.org/w/index.php?title=File:20091129_210544_FungusFromContourCabinAirFilter.jpg *License*: GNU Free Documentation License *Contributors*: User:Bob Blaylock

Image:20100207_234732_SeanetteSnotFungus.jpg *Source*: http://en.wikipedia.org/w/index.php?title=File:20100207_234732_SeanetteSnotFungus.jpg *License*: GNU Free Documentation License *Contributors*: Bob Blaylock at en.wikipedia

Image:20101121_212039_Mildew.jpg *Source*: http://en.wikipedia.org/w/index.php?title=File:20101121_212039_Mildew.jpg *License*: Creative Commons Attribution-Sharealike 3.0 *Contributors*: User:Bob Blaylock

Image:20100208_002523_ShowerCurtainMildew.jpg *Source*: http://en.wikipedia.org/w/index.php?title=File:20100208_002523_ShowerCurtainMildew.jpg *License*: GNU Free Documentation License *Contributors*: Bob Blaylock at en.wikipedia

Image:20100815_1818_Mold.jpg *Source*: http://en.wikipedia.org/w/index.php?title=File:20100815_1818_Mold.jpg *License*: GNU Free Documentation License *Contributors*: User:Bob Blaylock

Image:20100916_011605_Mycelium.jpg *Source*: http://en.wikipedia.org/w/index.php?title=File:20100916_011605_Mycelium.jpg *License*: GNU Free Documentation License *Contributors*: User:Bob Blaylock

Image:Blue Stilton Quarter Front.jpg *Source*: http://en.wikipedia.org/w/index.php?title=File:Blue_Stilton_Quarter_Front.jpg *License*: GNU Free Documentation License *Contributors*: User:Zerohund

Image:Moldytomato2500ppx.jpg *Source*: http://en.wikipedia.org/w/index.php?title=File:Moldytomato2500ppx.jpg *License*: Creative Commons Attribution-Sharealike 3.0 *Contributors*: User:Fastily

Image:Verschimmeltes Brot 2008-12-07.JPG *Source*: http://en.wikipedia.org/w/index.php?title=File:Verschimmeltes_Brot_2008-12-07.JPG *License*: Creative Commons Attribution-Sharealike 3.0 *Contributors*: User:Wimox

Image:Aspergillus_on_tomato.jpg *Source*: http://en.wikipedia.org/w/index.php?title=File:Aspergillus_on_tomato.jpg *License*: Creative Commons Attribution-Sharealike 3.0 *Contributors*: User:Multimotyl

Image:Coffee mold.JPG *Source*: http://en.wikipedia.org/w/index.php?title=File:Coffee_mold.JPG *License*: Public Domain *Contributors*: User:Aklibbee

Image:RiceMold.JPG *Source*: http://en.wikipedia.org/w/index.php?title=File:RiceMold.JPG *License*: GNU Free Documentation License *Contributors*: User:DerWaschbar

file:Fungi_collage.jpg *Source*: http://en.wikipedia.org/w/index.php?title=File:Fungi_collage.jpg *License*: Creative Commons Attribution-Sharealike 2.5 *Contributors*: BorgQueen

image:Omphalotus nidiformis Binnamittalong 2 email.jpg *Source*: http://en.wikipedia.org/w/index.php?title=File:Omphalotus_nidiformis_Binnamittalong_2_email.jpg *License*: Creative Commons Attribution-Sharealike 2.5 *Contributors*: Cas Liber

File:Penicillium labeled cropped.jpg *Source*: http://en.wikipedia.org/w/index.php?title=File:Penicillium_labeled_cropped.jpg *License*: Creative Commons Attribution-Sharealike 2.5 *Contributors*: User:Adrian J. Hunter, user:Y_tambe

Image:Armillaria ostoyae MO.jpg *Source*: http://en.wikipedia.org/w/index.php?title=File:Armillaria_ostoyae_MO.jpg *License*: Creative Commons Attribution-Sharealike 3.0 *Contributors*: Alan Rockefeller

File:DecayingPeachSmall.gif *Source*: http://en.wikipedia.org/w/index.php?title=File:DecayingPeachSmall.gif *License*: Creative Commons Attribution-Sharealike 2.0 *Contributors*: Andrew Dunn (User:Solipsist)

Image:Polyporus squamosus Molter.jpg *Source*: http://en.wikipedia.org/w/index.php?title=File:Polyporus_squamosus_Molter.jpg *License*: Creative Commons Attribution-Sharealike 3.0 *Contributors*: Dan Molter

Image:Morelasci.jpg *Source*: http://en.wikipedia.org/w/index.php?title=File:Morelasci.jpg *License*: Creative Commons Attribution 3.0 *Contributors*: User:Peter_G_Werner

Image:Cyathus stercoreus Fruchtkörper.JPG *Source*: http://en.wikipedia.org/w/index.php?title=File:Cyathus_stercoreus_Fruchtkörper.JPG *License*: Creative Commons Attribution-Sharealike 3.0 *Contributors*: Dr. Christian Glabasnia

Image:Arbuscular mycorrhiza microscope.jpg *Source*: http://en.wikipedia.org/w/index.php?title=File:Arbuscular_mycorrhiza_microscope.jpg *License*: Public Domain *Contributors*: Msturmel

Image:Ascocarp2.png *Source*: http://en.wikipedia.org/w/index.php?title=File:Ascocarp2.png *License*: GNU Free Documentation License *Contributors*: User:Debivort

Image:Neotyphodium coenophialum.jpg *Source*: http://en.wikipedia.org/w/index.php?title=File:Neotyphodium_coenophialum.jpg *License*: Public Domain *Contributors*: Nick Hill

Image:Lobaria pulmonaria 010108a.jpg *Source*: http://en.wikipedia.org/w/index.php?title=File:Lobaria_pulmonaria_010108a.jpg *License*: GNU Free Documentation License *Contributors*: User:BerndH

Image:Aecidium magnellanicum.jpg *Source*: http://en.wikipedia.org/w/index.php?title=File:Aecidium_magnellanicum.jpg *License*: Creative Commons Attribution-Sharealike 3.0 *Contributors*: Jason Hollinger

Image:S cerevisiae under DIC microscopy.jpg *Source*: http://en.wikipedia.org/w/index.php?title=File:S_cerevisiae_under_DIC_microscopy.jpg *License*: Public Domain *Contributors*: user:Masur

Image:Ganoderma lucidum 01.jpg *Source*: http://en.wikipedia.org/w/index.php?title=File:Ganoderma_lucidum_01.jpg *License*: Creative Commons Attribution-Sharealike 2.5 *Contributors*: Eric Steinert

Image:CordycepsSinensis.jpg *Source*: http://en.wikipedia.org/w/index.php?title=File:CordycepsSinensis.jpg *License*: Creative Commons Attribution-Sharealike 3.0 *Contributors*: User:Shyamal

File:Amanita phalloides 1.JPG *Source*: http://en.wikipedia.org/w/index.php?title=File:Amanita_phalloides_1.JPG *License*: GNU Free Documentation License *Contributors*: User:Archenzo

Image:Beauveria.jpg *Source*: http://en.wikipedia.org/w/index.php?title=File:Beauveria.jpg *License*: Public Domain *Contributors*: Stefan Jaronski

Image:Ergotamine3.png *Source*: http://en.wikipedia.org/w/index.php?title=File:Ergotamine3.png *License*: Public Domain *Contributors*: User:Elbreapoly

file:Adult deer tick(cropped).jpg *Source*: http://en.wikipedia.org/w/index.php?title=File:Adult_deer_tick(cropped).jpg *License*: Public Domain *Contributors*: Photo by Scott Bauer. (USDA ARS)

File:Ixodes scapularis.jpg *Source*: http://en.wikipedia.org/w/index.php?title=File:Ixodes_scapularis.jpg *License*: Public Domain *Contributors*: Content Provider(s) : CDC/ Michael L. Levin, Ph. D. Photo Credit : Jim Gathany

File:Tick engorged with thumb.jpg *Source*: http://en.wikipedia.org/w/index.php?title=File:Tick_engorged_with_thumb.jpg *License*: Creative Commons Attribution-Sharealike 2.5 *Contributors*: Original uploader was Jeremykemp at en.wikipedia

License

Creative Commons Attribution-ShareAlike 3.0 Unported - Deed

This is a human-readable summary of the Creative Commons Attribution ShareAlike 3.0 Unported License (http://en.wikipedia.org/wiki/Wikipedia:Text_of_Creative_Commons_Attribution-ShareAlike_3.0_Unported_License)
You are free:

- **to Share**—to copy, distribute and transmit the work, and
- **to Remix**—to adapt the work

Under the following conditions:

- **Attribution**—You must attribute the work in the manner specified by the author or licensor (but not in any way that suggests that they endorse you or your use of the work.)
- **Share Alike**—If you alter, transform, or build upon this work, you may distribute the resulting work only under the same, similar or a compatible license.

With the understanding that:

- **Waiver**—Any of the above conditions can be waived if you get permission from the copyright holder.
- **Other Rights**—In no way are any of the following rights affected by the license:
  - your fair dealing or fair use rights;
  - the author's moral rights; and
  - rights other persons may have either in the work itself or in how the work is used, such as publicity or privacy rights.
- **Notice**—For any reuse or distribution, you must make clear to others the license terms of this work. The best way to do that is with a link to http://creativecommons.org/licenses/by-sa/3.0/

GNU Free Documentation License

As of July 15, 2009 Wikipedia has moved to a dual-licensing system that supersedes the previous GFDL only licensing. In short, this means that text licensed under the GFDL can no longer be imported to Wikipedia. Additionally, text contributed after that date can not be exported under the GFDL license. See Wikipedia:Licensing update for further information.

Version 1.3, 3 November 2008 Copyright (C) 2000, 2001, 2002, 2007, 2008 Free Software Foundation, Inc. <http://fsf.org/>
Everyone is permitted to copy and distribute verbatim copies of this license document, but changing it is not allowed.

0. PREAMBLE

The purpose of this License is to make a manual, textbook, or other functional and useful document "free" in the sense of freedom: to assure everyone the effective freedom to copy and redistribute it, with or without modifying it, either commercially or noncommercially. Secondarily, this License preserves for the author and publisher a way to get credit for their work, while not being considered responsible for modifications made by others.
This License is a kind of "copyleft", which means that derivative works of the document must themselves be free in the same sense. It complements the GNU General Public License, which is a copyleft license designed for free software.
We have designed this License in order to use it for manuals for free software, because free software needs free documentation: a free program should come with manuals providing the same freedoms that the software does. But this License is not limited to software manuals; it can be used for any textual work, regardless of subject matter or whether it is published as a printed book. We recommend this License principally for works whose purpose is instruction or reference.

1. APPLICABILITY AND DEFINITIONS

This License applies to any manual or other work, in any medium, that contains a notice placed by the copyright holder saying it can be distributed under the terms of this License. Such a notice grants a world-wide, royalty-free license, unlimited in duration, to use that work under the conditions stated herein. The "Document", below, refers to any such manual or work. Any member of the public is a licensee, and is addressed as "you". You accept the license if you copy, modify or distribute the work in a way requiring permission under copyright law.
A "Modified Version" of the Document means any work containing the Document or a portion of it, either copied verbatim, or with modifications and/or translated into another language.
A "Secondary Section" is a named appendix or a front-matter section of the Document that deals exclusively with the relationship of the publishers or authors of the Document to the Document's overall subject (or to related matters) and contains nothing that could fall directly within that overall subject. (Thus, if the Document is in part a textbook of mathematics, a Secondary Section may not explain any mathematics.) The relationship could be a matter of historical connection with the subject or with related matters, or of legal, commercial, philosophical, ethical or political position regarding them.
The "Invariant Sections" are certain Secondary Sections whose titles are designated, as being those of Invariant Sections, in the notice that says that the Document is released under this License. If a section does not fit the above definition of Secondary then it is not allowed to be designated as Invariant. The Document may contain zero Invariant Sections. If the Document does not identify any Invariant Sections then there are none.
The "Cover Texts" are certain short passages of text that are listed, as Front-Cover Texts or Back-Cover Texts, in the notice that says that the Document is released under this License. A Front-Cover Text may be at most 5 words, and a Back-Cover Text may be at most 25 words.
A "Transparent" copy of the Document means a machine-readable copy, represented in a format whose specification is available to the general public, that is suitable for revising the document straightforwardly with generic text editors or (for images composed of pixels) generic paint programs or (for drawings) some widely available drawing editor, and that is suitable for input to text formatters or for automatic translation to a variety of formats suitable for input to text formatters. A copy made in an otherwise Transparent file format whose markup, or absence of markup, has been arranged to thwart or discourage subsequent modification by readers is not Transparent. An image format is not Transparent if used for any substantial amount of text. A copy that is not "Transparent" is called "Opaque".
Examples of suitable formats for Transparent copies include plain ASCII without markup, Texinfo input format, LaTeX input format, SGML or XML using a publicly available DTD, and standard-conforming simple HTML, PostScript or PDF designed for human modification. Examples of transparent image formats include PNG, XCF and JPG. Opaque formats include proprietary formats that can be read and edited only by proprietary word processors, SGML or XML for which the DTD and/or processing tools are not generally available, and the machine-generated HTML, PostScript or PDF produced by some word processors for output purposes only.
The "Title Page" means, for a printed book, the title page itself, plus such following pages as are needed to hold, legibly, the material this License requires to appear in the title page. For works in formats which do not have any title page as such, "Title Page" means the text near the most prominent appearance of the work's title, preceding the beginning of the body of the text.
The "publisher" means any person or entity that distributes copies of the Document to the public.
A section "Entitled XYZ" means a named subunit of the Document whose title either is precisely XYZ or contains XYZ in parentheses following text that translates XYZ in another language. (Here XYZ stands for a specific section name mentioned below, such as "Acknowledgements", "Dedications", "Endorsements", or "History".) To "Preserve the Title" of such a section when you modify the Document means that it remains a section "Entitled XYZ" according to this definition.
The Document may include Warranty Disclaimers next to the notice which states that this License applies to the Document. These Warranty Disclaimers are considered to be included by reference in this License, but only as regards disclaiming warranties: any other implication that these Warranty Disclaimers may have is void and has no effect on the meaning of this License.

2. VERBATIM COPYING

You may copy and distribute the Document in any medium, either commercially or noncommercially, provided that this License, the copyright notices, and the license notice saying this License applies to the Document are reproduced in all copies, and that you add no other conditions whatsoever to those of this License. You may not use technical measures to obstruct or control the reading or further copying of the copies you make or distribute. However, you may accept compensation in exchange for copies. If you distribute a large enough number of copies you must also follow the conditions in section 3.
You may also lend copies, under the same conditions stated above, and you may publicly display copies.

3. COPYING IN QUANTITY

If you publish printed copies (or copies in media that commonly have printed covers) of the Document, numbering more than 100, and the Document's license notice requires Cover Texts, you must enclose the copies in covers that carry, clearly and legibly, all these Cover Texts: Front-Cover Texts on the front cover, and Back-Cover Texts on the back cover. Both covers must also clearly and legibly identify you as the publisher of these copies. The front cover must present the full title with all words of the title equally prominent and visible. You may add other material on the covers in addition. Copying with changes limited to the covers, as long as they preserve the title of the Document and satisfy these conditions, can be treated as verbatim copying in other respects.
If the required texts for either cover are too voluminous to fit legibly, you should put the first ones listed (as many as fit reasonably) on the actual cover, and continue the rest onto adjacent pages.
If you publish or distribute Opaque copies of the Document numbering more than 100, you must either include a machine-readable Transparent copy along with each Opaque copy, or state in or with each Opaque copy a computer-network location from which the general network-using public has access to download using public-standard network protocols a complete Transparent copy of the Document, free of added material. If you use the latter option, you must take reasonably prudent steps, when you begin distribution of Opaque copies in quantity, to ensure that this Transparent copy will remain thus accessible at the stated location until at least one year after the last time you distribute an Opaque copy (directly or through your agents or retailers) of that edition to the public.
It is requested, but not required, that you contact the authors of the Document well before redistributing any large number of copies, to give them a chance to provide you with an updated version of the Document.

4. MODIFICATIONS

You may copy and distribute a Modified Version of the Document under the conditions of sections 2 and 3 above, provided that you release the Modified Version under precisely this License, with the Modified Version filling the role of the Document, thus licensing distribution and modification of the Modified Version to whoever possesses a copy of it. In addition, you must do these things in the Modified Version:

A. Use in the Title Page (and on the covers, if any) a title distinct from that of the Document, and from those of previous versions (which should, if there were any, be listed in the History section of the Document). You may use the same title as a previous version if the original publisher of that version gives permission.
B. List on the Title Page, as authors, one or more persons or entities responsible for authorship of the modifications in the Modified Version, together with at least five of the principal authors of the Document (all of its principal authors, if it has fewer than five), unless they release you from this requirement.
C. State on the Title page the name of the publisher of the Modified Version, as the publisher.
D. Preserve all the copyright notices of the Document.
E. Add an appropriate copyright notice for your modifications adjacent to the other copyright notices.
F. Include, immediately after the copyright notices, a license notice giving the public permission to use the Modified Version under the terms of this License, in the form shown in the Addendum below.
G. Preserve in that license notice the full lists of Invariant Sections and required Cover Texts given in the Document's license notice.
H. Include an unaltered copy of this License.
I. Preserve the section Entitled "History", Preserve its Title, and add to it an item stating at least the title, year, new authors, and publisher of the Modified Version as given on the Title Page. If there is no section Entitled "History" in the Document, create one stating the title, year, authors, and publisher of the Document as given on its Title Page, then add an item describing the Modified Version as stated in the previous sentence.
J. Preserve the network location, if any, given in the Document for public access to a Transparent copy of the Document, and likewise the network locations given in the Document for previous versions it was based on. These may be placed in the "History" section. You may omit a network location for a work that was published at least four years before the Document itself, or if the original publisher of the version it refers to gives permission.
K. For any section Entitled "Acknowledgements" or "Dedications", Preserve the Title of the section, and preserve in the section all the substance and tone of each of the contributor acknowledgements and/or dedications given therein.
L. Preserve all the Invariant Sections of the Document, unaltered in their text and in their titles. Section numbers or the equivalent are not considered part of the section titles.
M. Delete any section Entitled "Endorsements". Such a section may not be included in the Modified version.
N. Do not retitle any existing section to be Entitled "Endorsements" or to conflict in title with any Invariant Section.
O. Preserve any Warranty Disclaimers.

If the Modified Version includes new front-matter sections or appendices that qualify as Secondary Sections and contain no material copied from the Document, you may at your option designate some or all of these sections as invariant. To do this, add their titles to the list of Invariant Sections in the Modified Version's license notice. These titles must be distinct from any other section titles.
You may add a section Entitled "Endorsements", provided it contains nothing but endorsements of your Modified Version by various parties—for example, statements of peer review or that the text has been approved by an organization as the authoritative definition of a standard.
You may add a passage of up to five words as a Front-Cover Text, and a passage of up to 25 words as a Back-Cover Text, to the end of the list of Cover Texts in the Modified Version. Only one passage of Front-Cover Text and one of Back-Cover Text may be added by (or through arrangements made by) any one entity. If the Document already includes a cover text for the same cover, previously added by you or by arrangement made by the same entity you are acting on behalf of, you may not add another; but you may replace the old one, on explicit permission from the previous publisher that added the old one.
The author(s) and publisher(s) of the Document do not by this License give permission to use their names for publicity for or to assert or imply endorsement of any Modified Version.

5. COMBINING DOCUMENTS

You may combine the Document with other documents released under this License, under the terms defined in section 4 above for modified versions, provided that you include in the combination all of the Invariant Sections of all of the original documents, unmodified, and list them all as Invariant Sections of your combined work in its license notice, and that you preserve all their Warranty Disclaimers.
The combined work need only contain one copy of this License, and multiple identical Invariant Sections may be replaced with a single copy. If there are multiple Invariant Sections with the same name but different contents, make the title of each such section unique by adding at the end of it, in parentheses, the name of the original author or publisher of that section if known, or else a unique number. Make the same adjustment to the section titles in the list of Invariant Sections in the license notice of the combined work.
In the combination, you must combine any sections Entitled "History" in the various original documents, forming one section Entitled "History"; likewise combine any sections Entitled "Acknowledgements", and any sections Entitled "Dedications". You must delete all sections Entitled "Endorsements".

6. COLLECTIONS OF DOCUMENTS

You may make a collection consisting of the Document and other documents released under this License, and replace the individual copies of this License in the various documents with a single copy that is included in the collection, provided that you follow the rules of this License for verbatim copying of each of the documents in all other respects.
You may extract a single document from such a collection, and distribute it individually under this License, provided you insert a copy of this License into the extracted document, and follow this License in all other respects regarding verbatim copying of that document.

7. AGGREGATION WITH INDEPENDENT WORKS

A compilation of the Document or its derivatives with other separate and independent documents or works, in or on a volume of a storage or distribution medium, is called an "aggregate" if the copyright resulting from the compilation is not used to limit the legal rights of the compilation's users beyond what the individual works permit. When the Document is included in an aggregate, this License does not apply to the other works in the aggregate which are not themselves derivative works of the Document.

If the Cover Text requirement of section 3 is applicable to these copies of the Document, then if the Document is less than one half of the entire aggregate, the Document's Cover Texts may be placed on covers that bracket the Document within the aggregate, or the electronic equivalent of covers if the Document is in electronic form. Otherwise they must appear on printed covers that bracket the whole aggregate.

8. TRANSLATION

Translation is considered a kind of modification, so you may distribute translations of the Document under the terms of section 4. Replacing Invariant Sections with translations requires special permission from their copyright holders, but you may include translations of some or all Invariant Sections in addition to the original versions of these Invariant Sections. You may include a translation of this License, and all the license notices in the Document, and any Warranty Disclaimers, provided that you also include the original English version of this License and the original versions of those notices and disclaimers. In case of a disagreement between the translation and the original version of this License or a notice or disclaimer, the original version will prevail.

If a section in the Document is Entitled "Acknowledgements", "Dedications", or "History", the requirement (section 4) to Preserve its Title (section 1) will typically require changing the actual title.

9. TERMINATION

You may not copy, modify, sublicense, or distribute the Document except as expressly provided under this License. Any attempt otherwise to copy, modify, sublicense, or distribute it is void, and will automatically terminate your rights under this License.

However, if you cease all violation of this License, then your license from a particular copyright holder is reinstated (a) provisionally, unless and until the copyright holder explicitly and finally terminates your license, and (b) permanently, if the copyright holder fails to notify you of the violation by some reasonable means prior to 60 days after the cessation.

Moreover, your license from a particular copyright holder is reinstated permanently if the copyright holder notifies you of the violation by some reasonable means, this is the first time you have received notice of violation of this License (for any work) from that copyright holder, and you cure the violation prior to 30 days after your receipt of the notice.

Termination of your rights under this section does not terminate the licenses of parties who have received copies or rights from you under this License. If your rights have been terminated and not permanently reinstated, receipt of a copy of some or all of the same material does not give you any rights to use it.

10. FUTURE REVISIONS OF THIS LICENSE

The Free Software Foundation may publish new, revised versions of the GNU Free Documentation License from time to time. Such new versions will be similar in spirit to the present version, but may differ in detail to address new problems or concerns. See http://www.gnu.org/copyleft/.

Each version of the License is given a distinguishing version number. If the Document specifies that a particular numbered version of this License "or any later version" applies to it, you have the option of following the terms and conditions either of that specified version or of any later version that has been published (not as a draft) by the Free Software Foundation. If the Document does not specify a version number of this License, you may choose any version ever published (not as a draft) by the Free Software Foundation. If the Document specifies that a proxy can decide which future versions of this License can be used, that proxy's public statement of acceptance of a version permanently authorizes you to choose that version for the Document.

11. RELICENSING

"Massive Multiauthor Collaboration Site" (or "MMC Site") means any World Wide Web server that publishes copyrightable works and also provides prominent facilities for anybody to edit those works. A public wiki that anybody can edit is an example of such a server. A "Massive Multiauthor Collaboration" (or "MMC") contained in the site means any set of copyrightable works thus published on the MMC site.

"CC-BY-SA" means the Creative Commons Attribution-Share Alike 3.0 license published by Creative Commons Corporation, a not-for-profit corporation with a principal place of business in San Francisco, California, as well as future copyleft versions of that license published by that same organization.

"Incorporate" means to publish or republish a Document, in whole or in part, as part of another Document.

An MMC is "eligible for relicensing" if it is licensed under this License, and if all works that were first published under this License somewhere other than this MMC, and subsequently incorporated in whole or in part into the MMC, (1) had no cover texts or invariant sections, and (2) were thus incorporated prior to November 1, 2008.

The operator of an MMC Site may republish an MMC contained in the site under CC-BY-SA on the same site at any time before August 1, 2009, provided the MMC is eligible for relicensing.

How to use this License for your documents

To use this License in a document you have written, include a copy of the License in the document and put the following copyright and license notices just after the title page:

> Copyright (c) YEAR YOUR NAME.
>
> Permission is granted to copy, distribute and/or modify this document
>
> under the terms of the GNU Free Documentation License, Version 1.3
>
> or any later version published by the Free Software Foundation;
>
> with no Invariant Sections, no Front-Cover Texts, and no Back-Cover Texts.
>
> A copy of the license is included in the section entitled "GNU
>
> Free Documentation License".

If you have Invariant Sections, Front-Cover Texts and Back-Cover Texts, replace the "with...Texts." line with this:

> with the Invariant Sections being LIST THEIR TITLES, with the
>
> Front-Cover Texts being LIST, and with the Back-Cover Texts being LIST.

If you have Invariant Sections without Cover Texts, or some other combination of the three, merge those two alternatives to suit the situation.

If your document contains nontrivial examples of program code, we recommend releasing these examples in parallel under your choice of free software license, such as the GNU General Public License, to permit their use in free software.

GNU Free Documentation License Version 1.2, November 2002 Copyright (C) 2000,2001,2002 Free Software Foundation, Inc. 59 Temple Place, Suite 330, Boston, MA 02111-1307 USA Everyone is permitted to copy and distribute verbatim copies of this license document, but changing it is not allowed.

0. PREAMBLE

The purpose of this License is to make a manual, textbook, or other functional and useful document "free" in the sense of freedom: to assure everyone the effective freedom to copy and redistribute it, with or without modifying it, either commercially or noncommercially. Secondarily, this License preserves for the author and publisher a way to get credit for their work, while not being considered responsible for modifications made by others. This License is a kind of "copyleft", which means that derivative works of the document must themselves be free in the same sense. It complements the GNU General Public License, which is a copyleft license designed for free software. We have designed this License in order to use it for manuals for free software, because free software needs free documentation: a free program should come with manuals providing the same freedoms that the software does. But this License is not limited to software manuals; it can be used for any textual work, regardless of subject matter or whether it is published as a printed book. We recommend this License principally for works whose purpose is instruction or reference.

1. APPLICABILITY AND DEFINITIONS

This License applies to any manual or other work, in any medium, that contains a notice placed by the copyright holder saying it can be distributed under the terms of this License. Such a notice grants a world-wide, royalty-free license, unlimited in duration, to use that work under the conditions stated herein. The "Document", below, refers to any such manual or work. Any member of the public is a licensee, and is addressed as "you". You accept the license if you copy, modify or distribute the work in a way requiring permission under copyright law. A "Modified Version" of the Document means any work containing the Document or a portion of it, either copied verbatim, or with modifications and/or translated into another language. A "Secondary Section" is a named appendix or a front-matter section of the Document that deals exclusively with the relationship of the publishers or authors of the Document to the Document's overall subject (or to related matters) and contains nothing that could fall directly within that overall subject. (Thus, if the Document is in part a textbook of mathematics, a Secondary Section may not explain any mathematics.) The relationship could be a matter of historical connection with the subject or with related matters, or of legal, commercial, philosophical, ethical or political position regarding them. The "Invariant Sections" are certain Secondary Sections whose titles are designated, as being those of Invariant Sections, in the notice that says that the Document is released under this License. If a section does not fit the above definition of Secondary then it is not allowed to be designated as Invariant. The Document may contain zero Invariant Sections. If the Document does not identify any Invariant Sections then there are none. The "Cover Texts" are certain short passages of text that are listed, as Front-Cover Texts or Back-Cover Texts, in the notice that says that the Document is released under this License. A Front-Cover Text may be at most 5 words, and a Back-Cover Text may be at most 25 words. A "Transparent" copy of the Document means a machine-readable copy, represented in a format whose specification is available to the general public, that is suitable for revising the document straightforwardly with generic text editors or (for images composed of pixels) generic paint programs or (for drawings) some widely available drawing editor, and that is suitable for input to text formatters or for automatic translation to a variety of formats suitable for input to text formatters. A copy made in an otherwise Transparent file format whose markup, or absence of markup, has been arranged to thwart or discourage subsequent modification by readers is not Transparent. An image format is not Transparent if used for any substantial amount of text. A copy that is not "Transparent" is called "Opaque". Examples of suitable formats for Transparent copies include plain ASCII without markup, Texinfo input format, LaTeX input format, SGML or XML using a publicly available DTD, and standard-conforming simple HTML, PostScript or PDF designed for human modification. Examples of transparent image formats include PNG, XCF and JPG. Opaque formats include proprietary formats that can be read and edited only by proprietary word processors, SGML or XML for which the DTD and/or processing tools are not generally available, and the machine-generated HTML, PostScript or PDF produced by some word processors for output purposes only. The "Title Page" means, for a printed book, the title page itself, plus such following pages as are needed to hold, legibly, the material this License requires to appear in the title page. For works in formats which do not have any title page as such, "Title Page" means the text near the most prominent appearance of the work's title, preceding the beginning of the body of the text. A section "Entitled XYZ" means a named subunit of the Document whose title either is precisely XYZ or contains XYZ in parentheses following text that translates XYZ in another language. (Here XYZ stands for a specific section name mentioned below, such as "Acknowledgements", "Dedications", "Endorsements", or "History".) To "Preserve the Title" of such a section when you modify the Document means that it remains a section "Entitled XYZ" according to this definition. The Document may include Warranty Disclaimers next to the notice which states that this License applies to the Document. These Warranty Disclaimers are considered to be included by reference in this License, but only as regards disclaiming warranties: any other implication that these Warranty Disclaimers may have is void and has no effect on the meaning of this License.

2. VERBATIM COPYING

You may copy and distribute the Document in any medium, either commercially or noncommercially, provided that this License, the copyright notices, and the license notice saying this License applies to the Document are reproduced in all copies, and that you add no other conditions whatsoever to those of this License. You may not use technical measures to obstruct or control the reading or further copying of the copies you make or distribute. However, you may accept compensation in exchange for copies. If you distribute a large enough number of copies you must also follow the conditions in section 3. You may also lend copies, under the same conditions stated above, and you may publicly display copies.

3. COPYING IN QUANTITY

If you publish printed copies (or copies in media that commonly have printed covers) of the Document, numbering more than 100, and the Document's license notice requires Cover Texts, you must enclose the copies in covers that carry, clearly and legibly, all these Cover Texts: Front-Cover Texts on the front cover, and Back-Cover Texts on the back cover. Both covers must also clearly and legibly identify you as the publisher of these copies. The front cover must present the full title with all words of the title equally prominent and visible. You may add other material on the covers in addition. Copying with changes limited to the covers, as long as they preserve the title of the Document and satisfy these conditions, can be treated as verbatim copying in other respects. If the required texts for either cover are too voluminous to fit legibly, you should put the first ones listed (as many as fit reasonably) on the actual cover, and continue the rest onto adjacent pages. If you publish or distribute Opaque copies of the Document numbering more than 100, you must either include a machine-readable Transparent copy along with each Opaque copy, or state in or with each Opaque copy a computer-network location from which the general network-using public has access to download using public-standard network protocols a complete Transparent copy of the Document, free of added material. If you use the latter option, you must take reasonably prudent steps, when you begin distribution of Opaque copies in quantity, to ensure that this Transparent copy will remain thus accessible at the stated location until at least one year after the last time you distribute an Opaque copy (directly or through your agents or retailers) of that edition to the public. It is requested, but not required, that you contact the authors of the Document well before redistributing any large number of copies, to give them a chance to provide you with an updated version of the Document.

4. MODIFICATIONS

You may copy and distribute a Modified Version of the Document under the conditions of sections 2 and 3 above, provided that you release the Modified Version under precisely this License, with the Modified Version filling the role of the Document, thus licensing distribution and modification of the Modified Version to whoever possesses a copy of it. In addition, you must do these things in the Modified Version: A. Use in the Title Page (and on the covers, if any) a title distinct from that of the Document, and from those of previous versions (which should, if there were any, be listed in the History section of the Document). You may use the same title as a previous version if the original publisher of that version gives permission. B. List on the Title Page, as authors, one or more persons or entities responsible for authorship of the modifications in the Modified Version, together with at least five of the principal authors of the Document (all of its principal authors, if it has fewer than five), unless they release you from this requirement. C. State on the Title page the name of the publisher of the Modified Version, as the publisher. D. Preserve all the copyright notices of the Document. E. Add an appropriate copyright notice for your modifications adjacent to the other copyright notices. F. Include, immediately after the copyright notices, a license notice giving the public permission to use the Modified Version under the terms of this License, in the form shown in the Addendum below. G. Preserve in that license notice the full lists of Invariant Sections and required Cover Texts given in the Document's license notice. H. Include an unaltered copy of this License. I. Preserve the section Entitled "History", Preserve its Title, and add to it an item stating at least the title, year, new authors, and publisher of the Modified Version as given on the Title Page. If there is no section Entitled "History" in the Document, create one stating the title, year, authors, and publisher of the Document as given on its Title Page, then add an item describing the Modified Version as stated in the previous sentence. J. Preserve the network location, if any, given in the Document for public access to a Transparent copy of the Document, and likewise the network locations given in the Document for previous versions it was based on. These may be placed in the "History" section. You may omit a network location for a work that was published at least four years before the Document itself, or if the original publisher of the version it refers to gives permission. K. For any section Entitled "Acknowledgements" or "Dedications", Preserve the Title of the section, and preserve in the section all the substance and tone of each of the contributor acknowledgements and/or dedications given therein. L. Preserve all the Invariant Sections of the Document, unaltered in their text and in their titles. Section numbers or the equivalent are not considered part of the section titles. M. Delete any section Entitled "Endorsements". Such a section may not be included in the Modified Version. N. Do not retitle any existing section to be Entitled "Endorsements" or to conflict in title with any Invariant Section. O. Preserve any Warranty Disclaimers. If the Modified Version includes new front-matter sections or appendices that qualify as Secondary Sections and contain no material copied from the Document, you may at your option designate some or all of these sections as invariant. To do this, add their titles to the list of Invariant Sections in the Modified Version's license notice. These titles must be distinct from any other section titles. You may add a section Entitled "Endorsements", provided it contains nothing but endorsements of your Modified Version by various parties--for example, statements of peer review or that the text has been approved by an organization as the authoritative definition of a standard. You may add a passage of up to five words as a Front-Cover Text, and a passage of up to 25 words as a Back-Cover Text, to the end of the list of Cover Texts in the Modified Version. Only one passage of Front-Cover Text and one of Back-Cover Text may be added by (or through arrangements made by) any one entity. If the Document already includes a cover text for the same cover, previously added by you or by arrangement made by the same entity you are acting on behalf of, you may not add another; but you may replace the old one, on explicit permission from the previous publisher that added the old one. The author(s) and publisher(s) of the Document do not by this License give permission to use their names for publicity for or to assert or imply endorsement of any Modified Version.

5. COMBINING DOCUMENTS

You may combine the Document with other documents released under this License, under the terms defined in section 4 above for modified versions, provided that you include in the combination all of the Invariant Sections of all of the original documents, unmodified, and list them all as Invariant Sections of your combined work in its license notice, and that you preserve all their Warranty Disclaimers. The combined work need only contain one copy of this License, and multiple identical Invariant Sections may be replaced with a single copy. If there are multiple Invariant Sections with the same name but different contents, make the title of each such section unique by adding at the end of it, in parentheses, the name of the original author or publisher of that section if known, or else a unique number. Make the same adjustment to the section titles in the list of Invariant Sections in the license notice of the combined work. In the combination, you must combine any sections Entitled "History" in the various original documents, forming one section Entitled "History"; likewise combine any sections Entitled "Acknowledgements", and any sections Entitled "Dedications". You must delete all sections Entitled "Endorsements".

6. COLLECTIONS OF DOCUMENTS

You may make a collection consisting of the Document and other documents released under this License, and replace the individual copies of this License in the various documents with a single copy that is included in the collection, provided that you follow the rules of this License for verbatim copying of each of the documents in all other respects. You may extract a single document from such a collection, and distribute it individually under this License, provided you insert a copy of this License into the extracted document, and follow this License in all other respects regarding verbatim copying of that document.

7. AGGREGATION WITH INDEPENDENT WORKS

A compilation of the Document or its derivatives with other separate and independent documents or works, in or on a volume of a storage or distribution medium, is called an "aggregate" if the copyright resulting from the compilation is not used to limit the legal rights of the compilation's users beyond what the individual works permit. When the Document is included in an aggregate, this License does not apply to the other works in the aggregate which are not themselves derivative works of the Document. If the Cover Text requirement of section 3 is applicable to these copies of the Document, then if the Document is less than one half of the entire aggregate, the Document's Cover Texts may be placed on covers that bracket the Document within the aggregate, or the electronic equivalent of covers if the Document is in electronic form. Otherwise they must appear on printed covers that bracket the whole aggregate.

8. TRANSLATION

Translation is considered a kind of modification, so you may distribute translations of the Document under the terms of section 4. Replacing Invariant Sections with translations requires special permission from their copyright holders, but you may include translations of some or all Invariant Sections in addition to the original versions of these Invariant Sections. You may include a translation of this License, and all the license notices in the Document, and any Warranty Disclaimers, provided that you also include the original English version of this License and the original versions of those notices and disclaimers. In case of a disagreement between the translation and the original version of this License or a notice or disclaimer, the original version will prevail. If a section in the Document is Entitled "Acknowledgements", "Dedications", or "History", the requirement (section 4) to Preserve its Title (section 1) will typically require changing the actual title.

9. TERMINATION

You may not copy, modify, sublicense, or distribute the Document except as expressly provided for under this License. Any other attempt to copy, modify, sublicense or distribute the Document is void, and will automatically terminate your rights under this License. However, parties who have received copies, or rights, from you under this License will not have their licenses terminated so long as such parties remain in full compliance.

10. FUTURE REVISIONS OF THIS LICENSE

The Free Software Foundation may publish new, revised versions of the GNU Free Documentation License from time to time. Such new versions will be similar in spirit to the present version, but may differ in detail to address new problems or concerns. See http://www.gnu.org/copyleft/. Each version of the License is given a distinguishing version number. If the Document specifies that a particular numbered version of this License "or any later version" applies to it, you have the option of following the terms and conditions either of that specified version or of any later version that has been published (not as a draft) by the Free Software Foundation. If the Document does not specify a version number of this License, you may choose any version ever published (not as a draft) by the Free Software Foundation. ADDENDUM: How to use this License for your documents To use this License in a document you have written, include a copy of the License in the document and put the following copyright and license notices just after the title page: Copyright (c) YEAR YOUR NAME. Permission is granted to copy, distribute and/or modify this document under the terms of the GNU Free Documentation License, Version 1.2 or any later version published by the Free Software Foundation; with no Invariant Sections, no Front-Cover Texts, and no Back-Cover Texts. A copy of the license is included in the section entitled "GNU Free Documentation License". If you have Invariant Sections, Front-Cover Texts and Back-Cover Texts, replace the "with...Texts." line with this: with the Invariant Sections being LIST THEIR TITLES, with the Front-Cover Texts being LIST, and with the Back-Cover Texts being LIST. If you have Invariant Sections without Cover Texts, or some other combination of the three, merge those two alternatives to suit the situation. If your document contains nontrivial examples of program code, we recommend releasing these examples in parallel under your choice of free software license, such as the GNU General Public License, to permit their use in free software.

MIX
Papier aus verantwortungsvollen Quellen
Paper from responsible sources
FSC® C105338

Printed by Books on Demand GmbH, Norderstedt / Germany